Daniel S. [from old catalog] Curtiss

Wheat Culture. How to Double the Yield and Increase the Profits

Daniel S. [from old catalog] Curtiss

Wheat Culture. How to Double the Yield and Increase the Profits

ISBN/EAN: 9783743308749

Manufactured in Europe, USA, Canada, Australia, Japa

Cover: Foto ©berggeist007 / pixelio.de

Manufactured and distributed by brebook publishing software
(www.brebook.com)

Daniel S. [from old catalog] Curtiss

Wheat Culture. How to Double the Yield and Increase the Profits

WHEAT CULTURE.

HOW TO DOUBLE THE YIELD AND INCREASE THE PROFITS.

BY

D. S. CURTISS,

Washington, D.C.

ILLUSTRATED.

NEW YORK:

ORANGE JUDD COMPANY,

No. 245 Broadway.

NEW AMERICAN FARM BOOK.

ORIGINALLY BY

R. L. ALLEN,

AUTHOR OF "DISEASES OF DOMESTIC ANIMALS," AND FORMERLY EDITOR OF
THE "AMERICAN AGRICULTURIST."

REVISED AND ENLARGED BY

LEWIS F. ALLEN,

AUTHOR OF "AMERICAN CATTLE," EDITOR OF THE "AMERICAN SHORT-HORN
HERD BOOK," ETC.

CONTENTS:

INTRODUCTION.—Tillage Husbandry —Grazing — Feeding — Breeding — Planting, etc.

CHAPTER I.—Soils — Classification— Description — Management — Properties.

CHAPTER II.—Inorganic Manures— Mineral — Stone — Earth — Phosphatic.

CHAPTER III. — Organic Manures — Their Composition—Animal—Vegetable.

CHAPTER IV.—Irrigation and Draining.

CHAPTER V.—Mechanical Divisions of Soils — Spading — Plowing—Implements.

CHAPTER VI.—The Grasses—Clovers — Meadows — Pastures — Comparative Values of Grasses—Implements for their Cultivation.

CHAPTER VII.—Grain, and its Cultivation — Varieties — Growth—Harvesting.

CHAPTER VIII.—Leguminous Plants —The Pea—Bean — English Field Bean—Tare or Vetch—Cultivation —Harvesting.

CHAPTER IX.—Roots and Esculents— Varieties—Growth — Cultivation — Securing the Crops—Uses—Nutritive Equivalents of Different Kinds of Forage.

CHAPTER X.—Fruits—Apples—Cider —Vinegar—Pears—Quinces—Plums Peaches — Apricots — Nectarines— Smaller Fruits—Planting—Cultivation—Gathering—Preserving.

CHAPTER XI.—Miscellaneous Objects of Cultivation, aside from the Ordinary Farm Crops—Broom-corn— Flax—Cotton—Hemp—Sugar Cane Sorghum—Maple Sugar —Tobacco— Indigo—Madder—Wood—Sumach— Teasel — Mustard — Hops — Castor Bean.

CHAPTER XII.—Aids and Objects of Agriculture — Rotation of Crops, and their Effects—Weeds—Restora-

tion of Worn-out Soils—Fertilizing Barren Lands—Utility of Birds— Fences — Hedges — Farm Roads— Shade Trees—Wood Lands—Time of Cutting Timber—Tools—Agricultural Education of the Farmer.

CHAPTER XIII.—Farm Buildings— House — Barn—Sheds — Cisterns — Various other Outbuildings—Steaming Apparatus.

CHAPTER XIV.—Domestic Animals —Breeding—Anatomy—Respiration —Consumption of Food.

CHAPTER XV.—Neat or Horned Cattle Devons — Herefords—Ayreshires — Galloways — Short - horns — Alderneys or Jerseys—Dutch or Holstein —Management from Birth to Milking, Labor, or Slaughter.

CHAPTER XVI.—The Dairy—Milk— Butter—Cheese—Different Kinds— Manner of Working.

CHAPTER XVII.—Sheep—Merino— Saxon—South Down—The Long-wooled Breeds—Cotswold—Lincoln — Breeding — Management — Shepherd Dogs.

CHAPTER XVIII.—The Horse—Description of Different Breeds—Their Various Uses—Breeding—Management.

CHAPTER XIX.—The Ass—Mule — Comparative Labor of Working Animals.

CHAPTER XX.—Swine — Different Breeds—Breeding—Rearing—Fattening—Curing Pork and Hams.

CHAPTER XXI.—Poultry—Hens, or Barn-door Fowls — Turkey — Peacock—Guinea Hen—Goose—Duck —Honey Bees.

CHAPTER XXII.—Diseases of Animals—What Authority Shall We Adopt?—Sheep — Swine — Treatment and Breeding of Horses.

CHAPTER XXIII.—Conclusion—General Remarks — The Farmer who Lives by his Occupation—The Amateur Farmer—Sundry Useful Tables.

SENT POST-PAID, PRICE $2.50.

ORANGE JUDD COMPANY,

245 Broadway, New-York.

HOW TO DOUBLE THE YIELD AND INCREASE THE PROFITS.

BY

D. S. CURTISS,

WASHINGTON, D. C.

ILLUSTRATED.

NEW YORK:
ORANGE JUDD COMPANY,
245 BROADWAY.
1880.

CONTENTS.

CHAPTER I.—WHEAT CULTURE.

How to Increase the Yield—The Farmer's Capital—Cost of Raising
Wheat 9

CHAPTER II.—THE WHEAT PLANT.

Geographical History—Botanical Origin—Spring and Winter Wheat...11

CHAPTER III.—HOW TO OBTAIN A LARGE YIELD.

First—Underdraining. Second—Deep Cultivation. Third—Pulveriz-
ing of the Soil. Fourth—Alkalies and Soluble Silica. Fifth—
Clover and Pasture. Sixth—Selection and Preparation of the
Seed...14

CHAPTER IV.—INCIDENTAL REQUISITES TO A LARGE YIELD.

Top-Dressing—Insects and Diseases—The Average Yield Doubled—
Improved Drills and Wheat Hoes—Early Harvesting—Rust, its
Prevention—Experiments in Indiana—Experiments in England...18

CHAPTER V.—PLANTING OR SOWING WHEAT.

Time to Plant—Benefits of Early Planting—Proper Depth to Plant—
Germination of Seeds—Quantity of Seed to the Acre—Tools and
Implements.. 24

CHAPTER VI.—IMPORTANCE OF THE WHEAT CROP.

Commerce and Population—Various Statistics—Export of Wheat in
1870, and Since—English Wheat Growing Decreasing.............31

CHAPTER VII.—FLOUR THE FORM IN WHICH TO SELL WHEAT.

Milling Employs Many Persons—Value of Bran and Shorts—Profits of
Milling—Incidental Benefits—The Straw Not to be Sold..........36

CHAPTER VIII.—VARIETIES MOST GROWN IN THE UNITED STATES.

Varieties Preferred in Different States—Experiments in Missouri Agricultural College—Experiments in Massachusetts—Varieties Grown
in New York—Experiments in Pennsylvania—Varieties in Tennessee and Virginia—Three New Varieties—Some English Pedigree
Wheats...39

(III)

CHAPTER IX.—GREEN MANURING AND PLOWING.

Plowing-under Green Crops—Plowing Prairie Land, Present Way—
    Plowing in the Gulf States....................................50

CHAPTER X.—RECAPITULATION OF OPERATIONS.

Eight Important Matters—More Knowledge Needed.................52

CHAPTER XI.—EXAMPLES OF SUCCESSFUL WHEAT CULTURE.

Other Successful Examples—Yield and Product for Sixteen Years—
    Responses to My Circulars—Queries Contained in the Circulars—
    Table Giving Condensed Reports...............................56

CHAPTER XII.—EXTRACTS FROM LETTERS.....................62

CHAPTER XIII.—DISEASES AND INSECTS.

Rust and Smut—Winter-killing of Wheat—The Chinch Bug and Hes-
    sian Fly—Midges—Granary or Barn Weevil.....................64

CHAPTER XIV.—IMPROVED MACHINERY AND IMPLEMENTS......68

CHAPTER XV.—ANALYSES OF WHEAT AND STRAW.............70

CHAPTER XVI.—CONCLUSION...................................72

PUBLISHERS' ANNOUNCEMENT.

The importance of the wheat crop as a source of revenue to the country induces the publication of this book. The object of the farmer should be to increase the product, improve the quality, and decrease the cost per bushel. The Publishers put forth this volume in the confident assurance that it will aid every wheat grower to accomplish these ends, and thus add greatly to his own prosperity and to the wealth of the country.

(v)

WHEAT CULTURE.

CHAPTER I.

WHEAT CULTURE.

HOW TO INCREASE THE YIELD.

It is a well-known fact that the average yield of wheat in this country is absurdly small, being only about fourteen bushels per acre—not half what it should and might be in so new a country—and that the profits of growing it are correspondingly light. All this we have long noticed with regret, and that feeling has stimulated us to prepare this little work, hoping that the facts presented in it may, to some extent, aid the growers to produce better results, to secure larger yields, and thereby larger profits.

Whatever a man believes he can do, if it be proper and he desires to do it, he is very likely to do. It is to the interest of wheat growers to greatly increase their yield per acre, to even double the prevailing average yield, and thereby double their profits. We are well satisfied that this can be done, and it is our desire and aim to convince them that they can easily do it; then, with that faith, they will be sure to accomplish the result.

We believe that fuller knowledge and more thought among farmers generally will surely lead to higher achievements in their important work; that increased knowledge of the subject will secure increased yield, and

7

also, as a consequence, afford enlarged profits for their operations.

THE FARMER'S CAPITAL.

Each acre of land, with its necessary appurtenances, constitutes the farmer's fixed capital. The more he can produce from each acre, without exhausting his soil, the greater will be his interest on the investment. Labor, tools, seed, teams, and fertilizers, are the temporary capital, and this capital is continually consumed and worn out, requiring as continually to be replenished.

Exhausting or robbing the soil from year to year by improvident management, is equivalent to a man's expending or reducing his capital—the principal—instead of only the interest or income. All business men know this to be a ruinous performance, which will, sooner or later, result in bankruptcy.

If a farmer has ten acres of land it is so much invested capital, and if by judicious culture he obtains from it three hundred bushels of wheat each year, instead of only one hundred and fifty bushels, it is so much increased income for the capital invested, which is the value of the ten acres—say one hundred dollars per acre, making a capital of one thousand dollars.

COST OF RAISING WHEAT.

From various data it is safe to assume that, on the majority of farms throughout the country, the cost of raising and marketing the wheat crop is about ten dollars per acre, including taxes and interest on land and the wear and tear of tools. Reliable statistics for the past few years show that the average yield per acre has been about fourteen bushels, and the average price of wheat per bushel about one dollar, giving an income of about fourteen dollars per acre annually, and a profit of four dollars per acre above cost of production, allowing nothing for the straw and refuse, which are required by, or should be returned to, the soil to leave it in fair condition. This gives little over one-third profit on the cost of the crop.

But, as a business transaction, what per cent of interest does it afford on the fixed capital invested? It gives four per cent on the value of the land at one hundred dollars per acre; certainly rather less than active business men are generally contented with. It will do for large capitalists, millionaires, who have bank and stock investments, and who give no labor or toil to earn and secure their incomes, but is too small return for working men with only limited investments, of a few hundreds or thousands of permanent capital.

Now, suppose that by doubling the expense of production in labor and manure to twenty dollars per acre, and thereby the crop or yield is doubled to twenty-eight bushels per acre of wheat, and, as in the other case, the wheat is worth one dollar per bushel, the profits will be eight dollars per acre instead of four, and the interest on the capital will be eight dollars, or eight per cent, just double, without doubling the capital; a showing that will tell pleasantly on the prosperity of the operator. These calculations can be carried out to any extent and on any farm operation by any school-boy or the farmer's children. Suppose, for instance, a farm of one hundred acres, on which it is desired to raise one thousand bushels of wheat every year; at twenty bushels the acre, fifty acres would be required for the desired crop; but at forty bushels, which many obtain, only twenty-five acres would be required for one thousand bushels.

WHAT THE DEPARTMENT OF AGRICULTURE SHOULD DO.

The results—improved agriculture and increased yield of wheat—which this little work is endeavoring to bring about, should be a leading object and an important part

of the business of the Department of Agriculture. That
Department should, long before this, have adopted the
practice of sending thousands of circulars to intelligent
practical farmers in all wheat-growing portions of the
nation, submitting interrogatories and requests for answers, in order to obtain statements and reports of the
largest yield, and the average yield, per acre of wheat in
each locality, together with the details of the modes and
conditions under which large yields and poor yields were
produced, also the kind of seed and soil employed in
the operation, and then publish the replies.

Such reports and details would afford highly practical
and useful lessons, and aid others in obtaining higher results by such examples ; but probably we shall not have
such practical service from the Department of Agriculture very soon ; at least, not until the agricultural papers
everywhere speak out, and the farming community rise
up in their might and demand the appointment of an
earnest, honest, capable agriculturist to fill the important
position of Commissioner, one who is not a speculator,
seeking *eclat,* and who will have more regard for the best
interests of agriculture than for his own purse and noto-
riety. Such an official would make the Department a
benefit to the farmers.

Of the vast and vital importance of agriculture Mr.
William Saunders some time ago wrote : "At no time
in our nation's history, more than at the present, has there
been greater necessity for the encouragement by Government of this 'Art of Arts'—Agriculture—which is the
foundation of wealth and greatness; for to that source
we must look for the means of paying the national debt.
It is the fountain whence must flow that material aid
without which it is impossible for civilized peoples to
exist."

Nothing is truer than the above remark. The farmer
feeds all, and he pays most of the Government expenses ;

he is taxed, through the tariff laws, on everything he buys, to give gain and wealth to commercial and manufacturing classes.

CHAPTER II.

THE WHEAT PLANT.

GEOGRAPHICAL HISTORY.

Writers on the subject differ widely as to the original home of our great bread cereal, Wheat (*Triticum vulgare*). Some state it to be India; others Persia, and we find it frequently mentioned in the Holy Bible. The earliest recorded history of man shows it to have been among his breadstuffs, and it has flourished to a greater or less extent wherever civilized people have made their habitation. Of course all localities where this grain may possibly be grown are not equally favorable to it. To understand the best conditions for successfully growing wheat is of more importance at this time than to know precisely its original home, though knowing that fact is of some moment as indicating, to some degree, the most suitable conditions for greatest success in its cultivation.

It is stated, and generally understood, that wheat first came to the United States from Mexico, and that it was introduced into that country by Cortez, or during his administration. One of the beneficent provisions of Divine Providence in regard to wheat is that it will flourish, to some extent, in a wider range of country, climate and soil, than any other bread grain now in use, thereby rendering it the most valuable of all for the human race ; but possibly Oats (*Avena*) will flourish in a warmer cli-

mate, and Barley (*Hordeum*) in a colder, than our
wheats.

Botanical authors differ about as widely as do others as
to the origin or derivation of the wheat plant, *Triticum*.
Some of them maintain that wheat sprang from an in-
ferior grain or grass, and from that has been improved
by cultivation, up to the superior grain which we now
find it. Others contend that it was originally, and from
the beginning, a pure, absolute wheat, with all the characteristics that it now presents, with increased excellence
attained by cultivation, in some varieties, as is the case
with horses, where the thorough-bred specimens show
superior points to the common farm horse.

The former class contend that wheat is derived from
Ægilops ovata, a handsome grass, one to two feet high,
resembling wheat more than other grasses do, but more
like barley than wheat, and found in the countries bor-
dering on the Mediterranean and Adriatic Seas. It is
held that this grass simply, by good culture, has resulted
in what is now our wheat. But in writing this little
work it is not our aim or province to settle these disputed
questions, in which the doctors disagree.

Another beautiful characteristic of this chief of the
cereals is its wonderful susceptibility to modifying in-
fluences, resulting, under intelligent management of
growers, in the production of new varieties, adapted to
great differences of circumstances, and rewarding the
cultivators with grains suited to their peculiar situations
and necessities. Ten or a dozen species of *Triticum* are
mentioned by some writers; while others refer all our
cultivated wheats to a single species, with hundreds of
varieties.

Great changes in wheat are effected by two processes,
that of hybridizing, and what is called the pedigree sys-

tem : both modes have given valuable sorts of wheat. The pedigree system is best and most convenient ; it consists in selecting, from year to year, the best specimens, saving them for seed and planting them year after year.

SPRING AND WINTER WHEAT.

The great mass of the wheat grown in this country is the *Triticum vulgare*, which is divided into two sub-species or races—*T. hibernum*, Winter Wheat ; and *T. æstivum*, Spring Wheat. These are arranged in many groups, as the bald and bearded, the hard and soft, the white and red ; and still further subdivided as varieties which are known by texture and color of the kernel, the color and quality of the chaff or straw, and by many other characteristics which need not be enumerated here.

In regions where forests abound, and where heavy loam or clay lands exist, winter varieties of wheat are most suitable. For light, friable soils, like the prairies, where there is but little snow, and the soil is liable to be blown away, spring varieties succeed best, because, being planted in spring, they are not subject to be laid bare and destroyed by winter wind and frost. On moist lands, such as river-bottoms and alluvial formations, the rapid-growing, quick-ripening varieties (whether winter or spring) succeed best. Maturing in shorter time, they are more likely to escape rust and other calamities incident to such localities.

Yet, almost every natural land can, by proper management, be made a fair wheat soil. Under-draining, thorough pulverization, and a fair supply of vegetable manures, with ashes or lime, will render sand, gravel, or clay land a suitable soil for successful wheat-growing. But, first of all, it must be well drained and made fine and rich.

CHAPTER III.

HOW TO OBTAIN A LARGE YIELD.

Besides the minor details, there are six essential requisites for the production of uniformly large yields, per acre, of sound wheat.

FIRST—UNDER-DRAINING.

There must absolutely be a well-drained, deep, porous, warm subsoil, to the depth of at least two feet, with no stagnant water, in order that air and moisture may freely circulate through all parts of the earth to that depth, which will also allow the plant roots to run down and spread out easily for their necessary nourishment. Where the land is naturally of a loose texture, as gravel and sand, to a goodly depth, or with a gravelly sub-soil, the artificial drainage is less needed.

SECOND—DEEP CULTIVATION.

Deep cultivation by the sub-soil plow, is absolutely necessary, to the depth of at least twelve to fifteen inches, according to the nature of the land—whether porous or tenacious and hard—so as to enable the soil to retain moisture in a dry time, and to allow an excess to pass off readily in a wet season, as well as to allow the roots to have easy, wide range. Deep cultivation is, therefore, equally beneficial against the effects of drouth as against the drowning of the plants ; being loose and mellow to a goodly depth, moisture from below can freely rise to the surface when it is dry and hot, and heavy rains can readily sink down when they form surplus water on the surface. This operation does not require the raw sub-soil to be brought to the top.

Most of the advantages of sub-soil plowing and deep cultivation will be lost or not realized, and even injury be done, if the land be not also well under-drained to a considerable depth—two feet at least—because the deep plowing makes a basin of the land so plowed, where surplus water will settle and remain stagnant, unless there are sufficient drains at a lower depth than the plowing, to freely carry off all excess of water. But the drainage being ample, the land cannot well be broken too deeply for best results in wheat-growing. Let the sub-soil plowing be done so as not to bring much of the raw, stiff under-earth to the top at first, and the next year it will be first-rate soil for grain.

THIRD—PULVERIZATION OF THE SOIL.

Perfect pulverization, by fine plowing, harrowing, and rolling, is highly important, and will be productive of beneficial results, in giving large yields, and will preserve the fertility and strength of the land, by preparing the soil and putting it in that comminuted form in which the rootlets can absorb and appropriate a greater portion of the nutriment than when it is in a lumpy condition. In fact, the constituents of the soil cannot be brought into that state of solution in which they must be before plants can appropriate them, until the soil is made very fine. No part of the earth, no matter how rich it may be, is available for plant use, until it is very finely pulverized. Hence, much crushing, stirring, and culture is necessary.

FOURTH—ALKALI AND SOLUBLE SILICA.

There must be a liberal quantity of alkali and soluble silica in the soil, in order to enable it to produce a heavy crop of healthy wheat. Alkaline matters, such as potash and lime, must be in the soil, to operate with the air and moisture in dissolving all the required elements or ingredients, in order that they may be **taken** up in plant growth ; otherwise failure is certain. Liebig and other chemists and experimenters have proved that but small quantities of potash and silica are necessary, but that these small quantities are absolutely essential, as are moisture and air—those powerful solvents which reduce the constituents of the soil to a liquid state, so that plants can use them.

FIFTH—CLOVER AND PLASTER.

With the above preparation thoroughly made—that is, under-draining and sub-soiling—plaster, on clover plowed under in rank growth, and with the use of good seed wheat—a yield of thirty to forty bushels the acre of sound wheat will be the result, three years out of four, as surely as fifteen to twenty is from the ordinary farm operations. If the drainage be thoroughly done, and the sub-soiling well done, twelve to fifteen inches deep, the sub-soiling will not be required oftener than every four or five years, and the ordinary plowing need not be more than six or seven inches deep in the intermediate years, and for plowing under clover or other green crops, or any manure, the plowing need not be more than five or six inches deep, with mellow sub-soil.

In order not to bring raw sub-soil to the surface, it is best to cut the main furrow eight to ten inches deep with a large plow and stout team. Then follow in that furrow with a single horse and small, narrow plow, which will break the sub-earth four to six inches deeper and not quite so wide as the first furrow, and the next furrow will fall into and cover the small one, leaving the old surface soil still near the top. Most farmers know of and have used the small, sharp sub-soil plows made on purpose for that work. and to great advantage.

It is found to be a good plan to apply the alkalies—ashes, lime, potash, and salt, or whatever is used—to the ground just before sowing or planting the wheat, and then harrow them into the surface at the rate of ten to fifteen bushels of lime, or six to eight bushels of ashes or salt to the acre.

SIXTH—SELECTION AND PREPARATION OF THE SEED.

Proper selection and preparation of seed are all-essential in getting highest results in wheat growing. Seed should be perfectly ripe, gathered, thrashed and binned without the least wetting or moulding, and without being cracked or heated in the lightning thrashers; it should be perfectly screened and cleaned in the fanning mill. Farmers would, in the long run, be the gainers if they would each year gather with the grain cradle and thrash by hand with flail on a clean barn floor, sufficient wheat for seed, selecting the best growth in their fields, and letting it stand until perfectly ripe, taking that which seems to be earliest in ripening. When ready to plant, soak the seed six to ten hours in brine, and roll in plaster to dry it for the drill.

In regard to seeding with clover and grass, there are several modes and varieties, and differences of opinion among growers. Our own experience for several years in different States, on various soils, as well as considerable observation and reading, lead us to believe that Redtop is better than Timothy to seed with Clover, principally because it comes to maturity nearer the same time with the clover; and we think early spring is the best time to sow the clover, say on the last snows of the season, or during the first spring showers, or just before them, so that they will cover the seed into the soil and cause early germination; but we would sow the grass seed (Red-top, Timothy, or Orchard-grass) at the time of sowing the wheat, so that it may get a start in the fall.

CHAPTER IV.

INCIDENTAL REQUIREMENTS TO A LARGE YIELD.

PREPARING THE SEED-BED.

Incidental to the six essential points named, is the planting of the seed and the immediate preparation of the surface to receive it. The ground should be more thoroughly harrowed than some farmers do it, to level and fine it as completely as possible, but all farmers well know that the harrow will not crush the lumps, though it cuts some of them to pieces while it pushes others aside. The roller crushes and finely powders nearly all of the surface soil, making a fine seed-bed for the drill to run through and plant the seed, which it leaves in shallow gutters, lightly covered with small ridges each side. The ridges prevent the seed and young plant from being blown bare in high winds, and will also catch the snow and hold it to cover and shelter the wheat.

TOP-DRESSING—INSECTS AND DISEASES.

When the grain is well up in the fall, it will more than pay the cost to spread six or eight bushels of plaster to the acre on the crop, and after the frosts appear and the plants begin to be dormant, a dressing of four to six bushels of common salt, per acre, will be worth more than the outlay, not only by making the crop more luxuriant, but also by affording much security against injury by rust and insects. In the spring again, as soon as the ground is dry enough to allow of walking over it comfortably, a dressing of four or five bushels to the acre of fine lime will afford still further security against all insects or diseases. Sowing lime and plaster as a top-dressing, fall and spring, is needed for each crop, but the ten or fifteen bushels applied in preparing the soil will be sufficient if given once in three years.

THE AVERAGE YIELD DOUBLED.

We have no hesitation in saying that the system above marked out, if faithfully carried out for five years or longer, will as surely give all the growers who practice it more than double the average yield per acre of wheat, as the common practice gives that average. Every one who reads this can calculate the cost, and he will find that, although it will cost him less than one-half more per acre, it will as surely give him full double returns, and generally even more than double. Every farmer knows that it will cost very little, or no more, to cut and gather an acre which yields thirty bushels, than one that yields only fifteen. It costs no more to plant it, so that all the extra cost is in sub-soil plowing and top-dressing with the lime and plaster, and preparing the seed.

IMPROVED DRILLS AND WHEAT HOES.

But if the grower would still further increase his yield, and without proportionally increasing the expense, he can effect it by first using the improved drill points. These spread the seed-grain further apart than the ordinary drill, require less seed, distribute more evenly in the soil, and give the same quantity of plants more room to grow and receive air and light freely.

Also, let it be planted in drills wide apart (fourteen to sixteen inches), so that it may be hoed between the drills in fall and spring, with either hand-hoes or horse-hoes, which can be done by either running a corn-cultivator through it, or, better still, by the use of the new wheat hoe shown in figure 1.

Hoeing wheat is very much in favor by those who have practised it, and is said to largely increase the yield, and to generally give a better quality of grain. It is much practised in England and other parts of Europe, and has been adopted by some growers in this country, who uniformly acknowledge valuable results therefrom. Among other advantages claimed for it are these: it more than doubles the yield for a given quantity of land and seed by allowing much better tillering out; it keeps the land clean, any cockle or other weeds can readily be removed

Fig. 1.—THE WHEAT HOE AT WORK.

that may get into the rows of wheat; better opportunity is afforded to dislodge insects and to apply ashes, lime, plaster, sulphur, or other remedies, for diseases and insects; the grain is more pleasantly cut and gathered, giving twice the profit on every acre.

The engravings, figures 2 and 3, show the difference between wheat not hoed and that hoed.

EARLY HARVESTING.

One important operation to assure large profits from the wheat crop, is early harvesting, as soon as it is passing out of the milk into the dough state. This course is too little known or observed by the great majority of farmers, and, when better understood, will be more widely adopted. Five very important advantages, besides several lesser ones, are derived from harvesting the wheat crop thus early :

First—It is largely a preventive of injury by rust, as rust ceases to affect the grain as soon as it is cut, while

Fig. 2.—WHEAT IN CLOSE DRILLS, UNCULTIVATED.

the substance in the straw perfects the grain if cut in the milk state. Second—It gives more and heavier grain. Third—It gives more and better flour to the bushel, as all the time the grain stands, after the dough state, it makes bran at the expense of starch and flour. Fourth—It causes less waste by shelling and scattering while har-

Fig. 3.—WHEAT WIDE APART AND HOED—TILLERED OUT.

vesting and handling. Fifth—The harvesting can be sooner begun and out of the way, for other work, and is more pleasantly done, as the straw is tougher and softer to handle than when perfectly ripe. For flour and milling purposes, wheat cut early is the best, but the small quantity needed for seed should stand until perfectly ripe.

RUST—ITS PREVENTION.

A writer in the "Tecumseh (Mich.) Herald" communicates the following on the subject of early harvest of wheat :—"Rust in wheat is caused, among other things, by exhaustion in the soil of requisite mineral matters,

such as soluble silica, potash, and some others, which are
required to make stiff, bright, well glazed straw; and
this condition is aggravated, or rather operated upon, by
climatic changes, to produce fungi or rust. When the
straw is too tender and soft, lacking sufficient flinty or
glazed covering, which is the case when it grows too succulent with excess of nitrogenous and lack of mineral
matters, it is liable to be ruptured if suddenly struck by
the sun while damp. When this state of things occurs,
an immediate sprinkling of plaster or of lime has been
sometimes known to arrest the disease and prevent serious
diaster to the grain ; but when it occurs late enough to
find the grain advanced to the milk or dough state, im-
mediately cutting the grain will save it from injury by
the rust, and secure a crop of sound wheat with somewhat injured straw only."

EXPERIMENTS IN INDIANA.

He also quotes an early writer, in the agricultural
reports from Indiana, who gives the following facts in his
own experience :
    "He sowed three equal fields of similar quality of soil,
and same kind of seed, to wheat, in September. On the
twenty-fifth of June following, rust appeared in all three
fields ; the wheat was just in the dough state. On that
day he cut one of the fields ; the second day he cut another field, leaving them lying to cure in the swath, as
the grain was quite green, in the dough state. Four days
later he cut the third field, which, by this time, was
badly rusted. Upon thrashing and weighing the grain,
separately, of each field, he found that No. 1 (the first
cut) gave twelve bushels the acre of grain, weighing
fifty-six pounds the measured bushel ; No. 2 gave eight
bushels the acre, weighing forty-six pounds ; and No. 3
gave less than the seed sown, of poor grain."
    "In 1858, ten years later, rust made its appearance

again on his place, and he made another test of the
utility of early harvest, with three patches of wheat. The
third week in June, when rust struck all of his wheat, he
at once cut one field, while very green, just passing out
of the milk ; two days after he cut the second field ;
three days later still he cut the third, by which time the
rust was very bad. The early cut was left to cure in the
swath. He thrashed and weighed each parcel separately,
as in the former experiment. The first cut gave twenty-
five bushels the acre, weighing sixty-one pounds the
bushel ; the second lot only half as much, and weighing
fifty-six pounds the bushel ; and the third lot much
poorer than the second."

Here are instructive lessons in regard to early harvest
and rust.

EXPERIMENTS IN ENGLAND.

"An English farmer reports cutting three lots of
wheat at different stages of maturity—in the milk, in the
dough, and fully ripe. He thrashed separately, and had
one hundred pounds of each carefully ground and the
results weighed. The one hundred pounds of wheat, cut
in the milk, made seventy-five pounds of flour, eleven
pounds shorts, twelve pounds bran ; that cut in the
dough made eighty pounds flour, five pounds shorts,
thirteen pounds bran ; that cut fully ripe made seventy-
two pounds flour, eleven pounds shorts, fifteen pounds
bran ; two pounds lost by milling in each case. This
shows the dough state made most flour, and the ripest
made the least flour and most bran." Bran is made at
the expense of flour, by standing late.

Mr. Reid, of Indiana, reports to the Agricultural Department that he cut half of a fifty-acre field of Mediterranean wheat in the dough state ; the balance ten days
later. The first gave most bushels, and weighed sixty-
five pounds ; the last, less bushels, weighing only sixty
pounds ; the first also made more and better flour.

CHAPTER V.

PLANTING OR SOWING WHEAT.

TIME TO PLANT.

In this matter, as in most others connected with plant life, it is safe to take nature as a guide to a considerable extent. In most cases her ways and habits are the true ones ; and, in the operation of planting our grains, that guide is eminently correct, making due allowances for the changed conditions of artificial sowing. Hence early planting is the correct system, as nature usually plants the seed very soon after it is ripe and ready to fall from the parent plant. This would indicate that wheat should be planted as soon after becoming ripe as the soil can be made ready to receive the seed, after harvest and thrashing. There will be little danger of rust or insects, however early the grain may be sown, if the seed is well soaked in brine and dried in plaster or lime, if the land is well drained and deeply cultivated, and if, furthermore, the crop be liberally dressed with salt, lime, or plaster, in late autumn or early spring. There will, also, be little or no danger of too rank growth, or blasting, or shrinking, if the soil be well pulverized and deeply cultivated, with a fair supply of potash or lime to secure a sufficiency of soluble silica to make sound, healthy straw and chaff. With all the proper, natural conditions, early planting is surely the best—from August first to September fifteenth, according to locality.

On this point Mr. C. E. Thorne, of the Ohio University Farm, makes the following report of his experiments :

"A piece of bottom land, about ten rods wide by thirty long, was laid off in five equal strips, each two rods wide, and all sown with Clawson wheat—with seed at the same rate per acre—on the ninth, sixteenth, twenty-third, and thirtieth of September, and the seventh of October, 1878.

"The results were as follows: Strip sown September ninth yielded at the rate of thirty-three and one-fifth bushels per acre; strip sown September sixteenth yielded at the rate of thirty and three-tenths bushels per acre; strip sown September twenty-third yielded at the rate of twenty-six and two-fifths bushels per acre; strip sown September thirtieth yielded at the rate of thirty-two and seven-tenths bushels per acre, and strip sown October seventh yielded at the rate of twenty-six and one-fifth bushels per acre."

Here it will be seen that the seed sown in the last half of September yielded best.

BENEFITS OF EARLY PLANTING.

Some of the benefits of early planting are that it will secure a stronger growth of plants during autumn for enduring the winter, giving them more power to resist any calamity that may attack the crop, besides giving more time for tillering-out and making a good fruitful stool; and should any grower fear that his crop will make too stout a growth, he can feed it down or mow it off, either being preferable to having a slim, late crop. We find the majority of testimony among intelligent, observing experimenters, to be largely in favor of early planting, as early, at least, as the middle of September, while our own opinion, from many years' experience, is that even fifteen to twenty days earlier than that is preferable—say from the tenth of August to the first of September.

And when the great mass of farmers come to know and prize the many benefits of early harvest, they will also see the utility of uniformly planting earlier than is now the common custom; this will bring forward ear-

2

lier harvests, leaving time and room to make more perfect preparations for early planting. But with early
harvesting of the main crop, a portion of the largest and
finest of the grain, sufficient for their needed seed, should
be left standing to ripen perfectly, to be gathered by hand
with cradle or sickle, and then also thrashed by hand
with the flail.

Many more arguments or reasons could be given for
early planting or early harvesting, but space requires us
to be brief.

PROPER DEPTH TO PLANT.

In the matter of depth to plant, as in regard to time
of sowing, nature's methods may be considered, making
due allowance for attendant circumstances. Nature
drops the seed on the surface, then covers it very slightly
with only dust and light leaf-mould or straw and chaff
from the parent plant and surrounding litter to shelter
it from the sun-rays; she plants in the shade, where decaying matters cover and nourish until the plant is fairly
rooted, but she never plants deeply nor covers heavily.

Several circumstances must dictate the proper depth
for wheat in different localities, such as the kind of soil,
the degree of temperature and moisture, and the season
at which the planting is done; these and other conditions must, more or less, control the matter, so that no
invariable rule can be laid down for all situations and
periods, but much must be left to the judgment and skill
of the operator. In light, porous soils, that are rather
dry and warm, more depth of covering will be needed
than in heavy, moist lands. About one inch in the former and three-fourths of an inch in the latter will not be
far from right, as a general practice. A depth of not
less than three-fourths of an inch nor more than an inch
and a half are probably the extremes for wheat, to secure
the best results. Sandy and gravelly lands will admit of

deeper planting than heavier, clayey lands; but the light, friable soils of the Western prairies probably require the deepest covering of any in which wheat is grown, as that soil is more liable to be blown about by the winds, and there is generally less snow in winter to protect the crops from extremes. Then, in autumn, when the soil for some inches below the surface is warmer than in the spring, it will do to plant deeper than in the latter season.

A writer in the "New England Farmer" recommends a depth of not less than half an inch nor more than one inch. The "Michigan Farmer" favors a quarter to half an inch as giving the best results in most cases.

GERMINATION OF SEEDS.

Air, moisture, and warmth are all necessary to cause seeds to germinate and send up plants; they will "come up" sooner in warm than in cold soils; in those that are moist than in very dry; in loose, porous, than in stiff, hard soils. Experiments have shown that wheat planted at different depths came up as follows: At half an inch, in ten days; one inch, in twelve days; at two inches, in eighteen days; but in some cases of favorable warm conditions, wheat at those depths has been known to come up in six to four days, not usually, however, so soon. A temperature of soil and air from fifty to sixty degrees is favorable for wheat, though it will sprout and grow at several degrees both below and above that.

The "American Cultivator" gives the following useful tables:

"Frequent complaints are made that seeds do not germinate, and dealers in them are found fault with when, very generally, the fault lies in the improper manner in which people plant them. Many take no heed of the condition of the soil or of the depth at which the seed should be planted. The temperature and moisture also have a controlling influence. The temperature of germination, of the following seeds, is:

	Lowest. Degrees F.	Highest. Degrees F.	Most rapid. Degrees F.
Wheat	41	104	84
Barley	41	104	83
Pea	44	102	84
Corn	48	115	93
Bean	49	111	79
Squash	54	115	93

"Air-dried seeds will imbibe water of absorption completely in from forty-eight to seventy-two hours, in the following percentage:

Mustard	8	Buckwheat	47	Oats	60	Pea	107
Millet	25	Barley	49	Hemp	60	Clover	118
Corn	44	Turnips	51	Kidney beans	95	Beets	121
Wheat	45	Rye	58	Horse beans	104	White clover	127

Mr. S. N. Betts, in the "Michigan Homestead," gives the following interesting results of his experiments:

"The figures at the top of the table indicate the depth in inches at which the different samples were planted, and the figures at the left the time at which they came up, respectively. The other figures are the number of kernels that germinated in each forty:

		½	¾	1	2	3	4	5	6	7
February 2	6 A. M.	13	7							
February 3	9 A. M.	33	34	9	11					
February 4	11 A. M.	33	35	36	24	23				
February 5	9 A. M.			39	33	36	2			
February 6	10 A. M.						26			
February 7	11 A. M.				34		30	8	4	
February 8	9 A. M.				35		32	16	10	7
February 9	9 A. M.							19	17	11
February 11	9 A. M.							20	19	

"It will be seen that the seed planted one inch in depth gave the best returns. That planted respectively at three-quarters of an inch and two inches in depth yielded the same number of kernels. Seed planted three inches deep produced good, and that planted more than four inches very poor results."

QUANTITY OF SEED TO THE ACRE.

As in many other farm matters, there is diversity of opinion as to the quantity of seed it is best to sow, but judgment and circumstances must determine the point in different situations. Different preparation of both seed and soil will render more or less seed necessary ; climate and season have much to do with it ; kind of soil and variety of wheat, also, have a bearing upon the question. Wheats which tiller largely, like Clawson, Fultz, Gold Medal, etc., need less seed to the acre. Rich, fertile soil requires less than poor land. A long season and warm climate require less, as affording better conditions for spreading and growing ; fine, deep pulverization of the soil, which gives heavier growth to each plant, needs less seed, and well-cleaned, sound grain requires less seed than otherwise. Then more seed is required when sown in the spring than in the fall on the same land. Many circumstances enter into the determination of the question, so that careful discretion should be exercised by each grower for his own special case. The manner of planting, whether by drill or broadcast, and the style of drill used, make more or less seed necessary. If seed is well screened and brined, with all light, foul seed skimmed off, of course less will be necessary. From three to six pecks, per acre, is about right, as a general rule.

Broadcast sowing is hardly safe with less than six pecks to the acre of good seed, to secure full seeding to all parts of the ground, as some spots will get too much, and some will not be covered. With drill planting the seed is more evenly distributed, and more completely covered, with none too much in any one place ; hence less is needed. Some styles of drills distribute the seed better than others, some of them making four pecks necessary, while with the others three pecks will be suffi- cient. If every kernel were properly planted, and all perfectly distributed and germinated, even much less than the above quantity would be needed to fully seed the ground. Yet, if the planting be not done in the very best manner, to secure the growth of all the seed, we would recommend too much rather than too little— say six to eight pecks to the acre.

TOOLS AND IMPLEMENTS.

Every prudent farmer will buy the best and most substantial implements and tools, and those of the best pattern ; in the long run they are the most economical. The greater despatch of work and saving in labor will more than pay the extra price, in a single season, of a superior implement over a poor one. Often the loss of time and damage to crop, from hindrance by breakage of a flimsy tool, more than offsets the higher cost of a first class implement. Furthermore, the same good farmer will always take care of and shelter his tools and implements from the weather, when not in use, and not leave them out in the fields to be storm beaten.

CHAPTER VI.

IMPORTANCE OF THE WHEAT CROP.

COMMERCE AND POPULATION.

As an additional stimulus to our farmers to make efforts for greater yield in the production of wheat, we will call their attention to its great importance in the commercial and financial world.

Wheat is now the great sensation in commercial circles everywhere, and is the liveliest of all commodities in general trade. Especially to the United States is the matter one of great and growing importance, as many foreign countries are becoming more and more dependent upon us for their supply of breadstuffs ; and it is alike our duty and interest to supply them as fully as possible, and as cheaply as can be, consistent with fair returns for our labor.

Our room and area are almost unlimited and our facilities unbounded. Our soils and localities are numerous and diversified, while our climate embraces a wide and varied range, and generally of the most congenial character—reaching from ocean to ocean, and from the tropics to the frozen zone. It seems emphatically our mission to feed the Old World in its decline. It has been our grand privilege to give the Old World, even in our youth, an example of the best form of human government yet known to them. And now, before we are half grown, it is our privilege, and within our power, to furnish them with the very means to sustain their natural lives, and avert from them threatened starvation.

REPORTS BY LETTERS.

Many results reported in numerous letters received by
the author, for last year's harvest, show that the maximum yield, in many sections of many States, ranged
from thirty, forty, fifty, up to sixty-one bushels per acre,
under thorough, judicious culture ; and many reports,
gathered from other authentic sources, for several years
past, in different States, show that as high as fifty to
sixty bushels per acre have frequently been obtained. Is
it unreasonable, then, to claim that the great majority of
farmers can more than double the average yield of fourteen bushels, and make the average even as high as
thirty bushels the acre ? For instance, take the mean
between these maximum rates of forty to sixty bushels,
which is fifty bushels, and we believe it not a very hard
matter for the majority of wheat growers to obtain that
figure of fifty bushels the acre.

When farmers reflect that their productions have
really become the controlling commodities in the commercial world, they will understand that they cannot
become too intelligent in their business, nor too well
informed in regard to the markets and trade, where they
must sell and buy. Daniel Webster is reported to have
once said, in a speech, "that the time was not far distant
when American wheat would regulate the money and
exchanges of Europe and America," a prediction already
well-nigh fulfilment ; and a similar remark was recently
made by an English statesman, that "the breadstuffs of
America would soon control the exchanges and commerce
of the world," which is being realized by the farmers of
America already.

VARIOUS STATISTICS.

Different reports and estimates show that the total
wheat product of the United States, in 1878, was very
nearly four hundred and twenty million bushels, on

about thirty-one million acres of land, being nearly an average of fourteen bushels per acre. This quantity gave our people, for home consumption, two hundred million bushels, allowing five bushels *per capita* for the entire population, estimated in round numbers at forty millions, while the people of Europe have not more than three to four bushels a head for all the population. For seed, it likewise allowed us sixty million bushels for thirty-two million acres the succeeding season, which is the average, probably, sown that year, and then left about one hundred and forty million bushels surplus for exportation, which is the quantity shown by various statistics to have been exported by the time the crop of 1879 was ready to go forward ; and for the crop of the latter year we have even larger figures. The acreage harvested in 1879 was about thirty-two million acres, and the entire product was not far from four hundred and forty million bushels, showing a trifle less than an average of fourteen bushels per acre for the whole area sown, which is an absurdly small yield for a new country and lands, such as ours, and which ought to be, and easily can be, doubled, if the farmers will all adopt the best known methods, whereby they can likewise double their profits.

　　The Duke of Beaufort has made somewhat detailed estimates of the cost of the growing and transportation of wheat in America, and is very emphatic in his conclusions, saying : "As to the expense, I have no doubt but wheat can be raised in the United States and be landed at Liverpool, from the average of distance of shipping points on the coast of the United States, at a cost of four shillings per bushel, or thirty-two shillings per quarter," and then asks, "Can you compete with this price in England ? I say, certainly not." The Duke sums up his letter as follows : "The result of my consideration of the subject is this—that climate, steam transport by land and sea, with the labor question on both sides of the ocean, have made it out of the power of our agriculturists to compete with the growers of wheat in America, and that our farmers must turn their attention to better and cheaper modes of raising beef and mutton; distance, with the difficulty and expense of transporting live and dead meats, gives us an advantage over them that we will be wise to improve, rather than waste time and capital in trying the impossible task of competing with them in growing wheat, or we shall be driven out of the meat market also by the Americans."

From the "English Agricultural Gazette" we copy the following sensible remarks : "It is more than probable that the acreage of wheat sown here, for 1880, will be considerably less than for many years; farmers are disheartened as to wheat culture here; they have lost confidence in their climate, soil, and market; the advisability of growing less wheat has been advocated here for some years by many of our agricultural leaders, notably by Mr. Lawes, and it is not difficult to restrict the acreage of wheat in the Kingdom."

EXPORTS OF WHEAT IN 1850, AND SINCE.

In 1850, the United States exported wheat and flour (reducing the flour to its equivalent in bushels) eight million bushels; 1860, about eighteen million bushels; 1877, over fifty-seven million one hundred and fifty-two thousand bushels; 1878, over one hundred and thirty-four million three hundred thousand bushels; and in 1879, known and estimated above one hundred and sixty-one million four hundred thousand bushels; and the greater portion of this vast export, every year, went to Great Britain. In 1878, that country imported into her own borders some fifty-seven million five hundred thousand cwts. of grain, flour, and meal, of which forty-eight per cent, nearly half, were received from the United States.

ENGLISH WHEAT-GROWING DECREASING.

Another fact is auspicious to the prospects of the American farmer, which is—that the number of persons engaged in grain-growing in Great Britain is on the decrease. By reference to reports in English journals, it will be seen that the number of persons there engaged in wheat-growing in 1861 was one million eight hundred and thirty-three thousand two hundred and ninety-five ; but in 1871 the number was decreased to one million six hundred and thirty-four thousand one hundred and ninety-two, a reduction of nearly twelve per cent in ten years ; and the decrease, during the past decade, is reported as being still larger, though the number engaged in grazing has remained as usual.

It is also reported that the number of acres sown to grain, especially to wheat, is steadily becoming less, for the past ten years.

Great Britain will, undoubtedly, for a long time, be the largest purchaser of our farm products, especially of wheat, while some other countries of Europe and of South America will often need portions of our grains, but they will want it mostly as flour, which is really the true form in which we should sell all of our surplus wheat.

From numerous reports and other sources, in foreign journals, we learn that the average yield, per acre, in France and Germany, until the last few years, was twenty-eight to thirty-two bushels ; and in England and Wales, from thirty to thirty-four bushels, until the late disastrous crops ; but that was the average yield, for many generations, even on their old lands, which had been cropped for ages.

CHAPTER VII.

FLOUR THE FORM IN WHICH TO SELL WHEAT.

MILLING EMPLOYS MANY PERSONS.

For several important reasons, all of our surplus wheat should be sold or exported in the shape of flour.

First—It will afford useful employment to a large number of mechanics and others laborers here at home, such as builders and operators of mills, . coopers, and others, in making barrels or other packages—in packing and putting up the packages, handling and hauling, besides other incidental labor, not required in selling and shipping whole wheat. The business and profits of feeding, clothing, housing, and otherwise maintaining all of these various operatives, inside and outside of the flouring mills, are likewise very considerable.

THE VALUE OF BRAN AND SHORTS.

Second—It will retain here at home the bran, shorts, and other refuse, always produced in milling, whence it can and always should go back to the farms and land where wheat is produced, as fertilizers to the soil, through feeding stock, to aid in preventing exhaustion or " running down " of the soil. It is well known, and is shown by various analyses, that the bran and straw contain nearly all of the mineral or inorganic matter which the wheat crop has derived from the soil. Consequently those portions of the wheat plant do most towards impoverishing the land and rendering it less capable of producing a heavy crop of sound grain ; hence as much as possible of the bran and straw should go back to the land.

THE PROFITS OF MILLING.

THIRD—The large profits of milling and making and packing flour, by which many large fortunes are acquired, will be retained and accumulated at home, affording attractive investments for a large amount of capital. Of the one hundred and forty million bushels surplus of 1878, perhaps as much as eighty million to one hundred million bushels were exported in the shape of whole wheat; that would make about twenty-five million barrels of flour, and at a casual guess it is safe to say that, including bran and shorts, the profits on milling that quantity of wheat would be one dollar per barrel, which would make the snug sum of twenty-five million dollars saved at home by grinding all of it into flour before exporting; no matter whether the figures are precisely correct or not, they illustrate the proposition and point the argument all the same.

INCIDENTAL BENEFITS.

FOURTH—Considerable saving in freights and insurance would be made, and less trouble in handling, as a mass of wheat, when reduced to the shape of well-packed flour, occupies less room, is liable to less risk, and can be more pleasantly handled than its equivalent as whole wheat. For instance, twenty-five million barrels of flour will not cost as much freight and insurance for transport from Chicago to New York, or from Baltimore to Liverpool, as would the quantity of wheat, one hundred million bushels, required to make it; consequently, the difference would be so much saving to be added to the profits at the point of shipping or milling. For these and other reasons, as much as possible of wheat should be made into flour before exporting, or even before being sent from the county where grown.

FIFTH—Where large flouring and coopering operations are carried on, many laborers of different classes are employed. They, in turn, aid the prosperity of the gardeners, orchardists, and small farmers, by consuming and making market for their vegetables, milk, fruits, and poultry products, to a considerable extent, upon which, generally, better profits are realized than on their wheat. Hence the agricultural classes should do what they can toward the building of mills in their neighborhoods, which will flour all of their surplus wheat before it leaves the vicinity where it is raised; and then the farmers should seek to get back to their own premises as much of the bran and shorts as they well can, to feed the stock and soil.

THE STRAW NOT TO BE SOLD.

It is certainly bad policy to sell the straw off of the farm, as it largely contains the soluble silica of the soil, which is so essential to make a vigorous, healthy crop of wheat. There are of late so many ways for using up straw, in making coarse paper and other fabrics, in towns and cities, which give it a merchantable price that offers tempting inducements for farmers to haul it to town for sale, in many districts, to the injury of their lands, by robbing them of their silica, without an adequate return. This in the long run will prove ruinous, unless an equivalent of useful manure of some kind is carried back and supplied to the soil. Nothing is really an equal substitute for straw except good stable manure, swamp muck, and leaf mould.

CHAPTER VIII.

VARIETIES MOST GROWN IN THE UNITED STATES.

The following is a list of varieties that have succeeded in most of the States, and proved to be superior in some desirable quality—either for earliness, hardiness, prolific yield, freedom from disease, or some other good characteristic, or for adaptability to certain localities :

Early May, Bald Mediterranean, Canada Flint, Velvet, Genesee Flint, Hutchinson, Kentucky, Indiana, New York Flint, Bearded Mediterranean, Turkish Flint, Harmon's White (New York Flint), Blue-stem, Boone, Gander, Hoover, Lambert, Michigan, Malta, Orange, Perkey, Golden-chaff, Quaker, Shot-berry or Starbuck, Dayton, Carolina, Golden-straw, Virginia, Reed-straw, Boughton or Tappahannock, Tennessee, Bald Genesee, and Zimmerman. The Early May, known also as Alabama, Early Ripe, June, and Watkins, has been cut as early as May twenty-sixth, in Ohio, yielded well, and weighed sixty-five pounds to the bushel. Mr. Klippart reports that the Orange has been known to yield seventy bushels to the acre, and eighty kernels in a single head ; and that the Early May, Genesee Flint, and Harmon's White, frequently weighed sixty-four to sixty-six pounds the bushel.

Among later varieties, which are gaining popularity as prolific yielders, are the Keystone, Amber, Red Mediterranean, and Yellow Missouri (winter), and Champlain, Defiance, Russian White, and Touce (spring) ; the heads of some of them are said to be eight inches long, with seventy to eighty kernels in them.

VARIETIES PREFERRED IN DIFFERENT STATES.

In Colorado, spring wheats mostly prevail, the White Australian proving very prolific. In Connecticut, Red

Winter, and Gold Medal, with the Sherman as a spring
wheat, have given good results. Delaware produces the
Virginia White and Fultz, and most other varieties of
winter wheats that succeed in Maryland. Illinois and
Iowa grow most of the winter and spring sorts that succeed in Wisconsin and other States generally, including
Fultz and Club. In Maryland, the Boughton, Blue-
stem, Clawson, Fultz, Gold Dust, Gold Medal, Jennings,
Lancaster, Mediterranean, and New York Flints, are
popular. In Michigan both spring and winter varieties
are grown extensively; of the latter, Clawson, Deihl,
Early May, Gold Medal, Genesee Flint, Lancaster,
Mediterranean, and Victor seem to be most popular ; of
the former, Arnautka, Canada Club, Champlain, Defiance, Fife, Milwaukee, and Touzelle are preferred.
Minnesota grows largely of Arnautka, Fife, Odessa, and
Club spring wheats and some winter sorts. Kansas
grows spring and some winter wheats.

EXPERIMENTS AT THE MISSOURI AGRICULTURAL COLLEGE.

In Missouri all the popular sorts succeed, particularly
Clawson and Sandford. Prof. G. C. Swallow, Dean of the
Agricultural College, writing in regard to some interesting experiments made with wheat on the farm of that
Institution, in 1877–78, reports that of sixty-one varieties of winter wheat experimented with, twelve were
winter-killed and one was destroyed by rust. Of the
remaining forty-eight kinds, all planted September
twenty-ninth, 1877, forty-three were harvested in June,
and five in July ; eight kinds grew to a hight of six feet ;
six kinds weighed the standard of sixty pounds, or over ;
five reached thirty bushels, or over, per acre ; two, less
than one hundred pounds of straw per bushel, namely :
Clawson, giving on an acre two thousand six hundred
and forty-six pounds of straw to twenty-eight bushels of
grain ; and the Sandford, giving on an acre one thousand

two hundred and fifty-two pounds of straw and eighteen and three-quarters bushels of grain.

The Missouri Agricultural College reports Red May winter wheat, the earliest ripening variety, raised on their experimental farm ; sowed September twenty-ninth, it was ripe on June eighth ; is a smooth, or beardless wheat ; gives about twenty-eight bushels the acre, weighing fifty-nine pounds.

The heaviest wheat which they raised was the Mediterranean, sixty-one pounds, and twenty-two bushels the acre—red grain and bearded heads. The largest yield of any was from Rogers' White, thirty-eight and three-quarters bushels per acre, very plump, weighing fifty-nine pounds.

EXPERIMENTS IN MASSACHUSETTS.

In Massachusetts, as reported, some years ago, by J. H. Klippart, from "Philosophic Transactions," it is stated that "C. Miller, of Cambridge, on June second, planted a few grains of red wheat ; one plant tillered out so much by August eighth that he was enabled to divide it into eighteen parts, all of which he planted separately in pots of earth. Then, in September and October, so many of these had multiplied their stalks that the number of plants was sixty-seven, which were divided and again set out separately. With the first growth of spring the tillering still went on, so that at the beginning of April a further division was made, and the number of plants was five hundred. These all proved to be extremely vigorous, more so than wheat plants under ordinary circumstances, so that the whole number of heads of wheat gathered from the original plant, by this process of division, was twenty-one thousand one hundred and nine. In a few instances there were one hundred heads on a single plant, very fine and long, some being seven inches in length and containing seventy grains

each. The grain, when all separated from the straw,
weighed forty-seven pounds and seven ounces, measuring
three pecks and six quarts, estimated number of grains
being five hundred and seventy-six thousand eight hundred and forty, and all from one grain in one harvest."

Mr. Killibrew remarks :—"Of course, such an enor-
mous increase is not practicable on a large scale. Yet the
experiment is useful as showing the vast power of increase possessed by this most valuable grain, under skillful, intelligent management, and is an encouragement
to our farmers to put forth their best efforts."

VARIETIES GROWN IN NEW YORK.

Probably, in New York, a greater number of varieties
of wheat are grown than in any other one State, possibly
equalled by Ohio and Pennsylvania, where pretty much
the same varieties are the general favorites. So far as we
have been able to learn, Boughton, Clawson, Deihl, Genesee Flint, Fultz, Wicks, Gold Dust, and Harmon's
White, are most popular, with Gold Medal, Jennings'
White, Mediterranean, and Early May, with some others,
are nearly as much so, all giving satisfactory results in
various localities. New York has long been distinguished
for its fine wheat and excellent flour ; the old, long time
ago popular " Genesee White Flint," known the world
over for the superior flour made from it, was of Spanish
origin, and has a wide progeny through the whole country—the Boughton, Tappahannock, Blue-stem, Harmon's
White, and many others, having originated from it.

Hon. L. L. Polk, Commissioner of Agriculture for
North Carolina, reports that the Fultz does well in that
State ; others report the Clawson as popular.

VARIETIES AND EXPERIMENTS IN OHIO.

For Ohio, Prof. C. E. Thorne reports, in the "Farm
and Friend," that "the wheat harvest commenced on the

twenty-fifth of June with the Velvet-chaff variety, a
hard, amber wheat, and is valued for its freedom from
disease, stiffness of straw, earliness, and good flour.
Fultz ripened about the same time, possessing good
qualities, with rather softer grain.   Golden Straw was
cut on the twenty-seventh.   It is a white, plump wheat,
originated in Tennessee, has short, stiff straw, but has
not proved a very heavy cropper.   June thirtieth Clawson was cut, and has sustained its high reputation for
freedom from disease, weight of crop, and good straw.
Next Gold Medal was cut, a soft, white grain, short,
stiff, clean straw, and heavy cropper, but shells easily.
About the same time the Silver-chaff was ripened, a
Canada wheat, is a tall grower, with stiff straw, not very
liable to lodge on any soil, appears free from disease,
does not shell easily, is white as the Clawson and flinty
as the Mediterranean.   Though not accurately measured,
the yield was about as follows :   Velvet-chaff, thirty-
six bushels per acre ; Fultz, twenty-six ; Golden Straw,
twenty-seven ; Clawson, thirty-four ; Gold Medal, thirty-
six, and Silver-chaff, thirty-five."

In the palmy days of the Genesee Flint, the splendid
varieties of Clawson, Fultz, Gold Dust, Gold Medal, and
Jennings' White, seem not to have been known, at least,
are not mentioned by Mr. Klippart in his work, though
they are now, perhaps, the five most popular varieties of
winter wheat grown in our country.   He names Canada
Flint, Genesee Flint, Hutchinson, English, Blue-stem,
Lambert, Orange, and Early May, as among the most
popular white wheats in 1860.

EXPERIMENTS IN PENNSYLVANIA.

Reports from the "Experimental Farm" of the Agri-
cultural College of Pennsylvania, of which Prof. James
Calder is President, show the Clawson, Gold Medal, Gold
Dust, Fultz, and Lancaster, to be the most desirable

varieties, among many, grown on their place, and perhaps throughout the State. The proportion of grain to straw is an important consideration in determining the value of any variety of wheat. We here give some important reports on the subject from the Pennsylvania Agricultural College and "Experimental Farm." Their experiments in 1878 included above twenty varieties, but I here give the results of the four most important varieties, viz. : Clawson, Fultz, Gold Dust and Gold Medal. They were all sown on Spetember twenty-eighth, 1877, and all harvested June twenty-eighth, 1878, with the same care and accuracy.

Fultz and Gold Medal, light amber and beardless, yielded, of grain and straw, per acre, as follows : Fultz—grain, thirty-two and eight one-hundreths bushels ; straw, two thousand five hundred and ninety-two pounds. Gold Medal—grain, thirty-one and fifty-four one-hundredths bushels ; straw, two thousand five hundred and fifty-two pounds, a remarkable nearness of yield, in both grain and straw, by these sorts.

Clawson and Gold Dust, beardless, whiter than above kinds, sown and harvested the same date as above, gave the following results : Clawson—grain, thirty-two bushels ; straw, three thousand and seventy-two pounds. Gold Dust—grain, thirty-one and twenty-four one-hundreths bushels; straw, three thousand and forty-two pounds; very nearly the same yields of straw and grain, by each, respectively and proportionally; but it will be noticed that the Fultz and Gold Medal gave slightly larger proportion of grain to straw than the Clawson and Gold Dust ; all of the other kinds (of the twenty tried) gave considerably more straw, compared to quantity of grain, than these four thus particularly mentioned.

The ground on which all of these were sown was a clayey, sandy loam wheat stubble, plowed soon after harvest, then liberally manured. The wheat was put in with the drill after the land was thoroughly rolled and all lumps crushed and powdered.

Genesee Flint, Boughton, Mediterranean, Silver-chaff, Blue-stem, Jennings' White, Victor, and Wicks, are some of the varieties which give large proportion of grain to straw, while Sandford, Eureka, "Bill Dallas," Walker, and Deihl, give greater proportion of straw to grain than those named above.

Farmers desiring a wheat which will produce the best proportion of grain to straw, will find a lesson and a guide in this statement.

VARIETIES IN TENNESSEE AND VIRGINIA.

Hon. J. B. Killebrew, in his instructive work on Wheats in Tennessee, mentions, as succeeding generally in that State, the Amber, Boughton, Clawson, Deihl, Early May, Fultz, Genesee Flint, Golden-straw, Lancaster, Mediterranean, Quaker, Walker, and some others; and he remarks, specially, that "before the introduction of Boughton, Clawson, Fultz, and Mediterranean, with some others, fifteen to twenty bushels the acre was considered an extra yield, but since then twenty-five to thirty-five bushels the acre are not uncommon on properly tilled lands."

Hon. Thomas Pollard, Commissioner of Agriculture for Virginia, in his excellent Report, 1879, shows that the varieties most grown and popular in that State are, about in the order named, the following: Fultz, Lancaster, Scott, Amber, Blue-stem, Clawson, Canada, Golden-straw, Mediterranean, and Genesee Flint; and others, popular in localities, as the Jennings' White, Michigan, Kentucky, Missouri Yellow, New York Flint, Quaker, Ruffin, Weeks, and Zimmerman.

Prof. J. R. Page, of the Virginia University Experimental Farm, in 1878, reports experiments with "Eureka" and "Fultz" wheats, planted with drill, one-half acre each, on November second, and harvested on June eleventh and fifteenth. Fultz, first cut, gave nine bushels the half acre, and eight hundred and thirty pounds of straw, grain weighing sixty pounds the bushel. Eureka gave eight and one-half bushels the half acre of grain, and one thousand one hundred and ninety-six pounds of straw; the grain weighed sixty-two pounds the bushel; the land was a gray, micaceous, sandy loam. He further experimented on six lots of land, of an acre each—poor, micaceous, siliceous soil, with many white flint rocks scattered over it. The land was all plowed and fallowed from the tenth to the eighteenth of September, 1877, harrowed, and wheat sowed—the Fultz by drill—October fifteenth; in the following March, was harrowed and sowed with clover seed. The wheat was harvested June tenth and eleventh. Lot one was manured with two hundred pounds ground bone, one hundred pounds nitrate of soda, one hundred pounds muriate of potash, in all, value nine dollars and seventy-five cents. Yield of grain was eighteen and one-half bushels, weighing sixty-one and one-half pounds per bushel; straw, one thousand three hundred and twenty-five pounds, and chaff, two hundred and seventy-nine pounds. Same lot, succeeding year, without fertilizers, produced two and one-half bushels grain, weighing sixty and one-half pounds per bushel, four hundred and twenty pounds straw, and thirty-three pounds chaff. Quantity of seed sown was five pecks per acre. The other five lots, treated in similar manner the same two years, gave similar results, less bushels of grain and of lighter weight.

"A seventh acre-lot was manured with two hundred pounds dissolved bone, one hundred pounds nitrate of soda, one hundred pounds potash, all mixed; value, nine dollars and seventy-five cents; sowed by drill with five pecks Fultz wheat, October fifteenth. Yield was eighteen bushels grain, weighing sixty-one pounds per

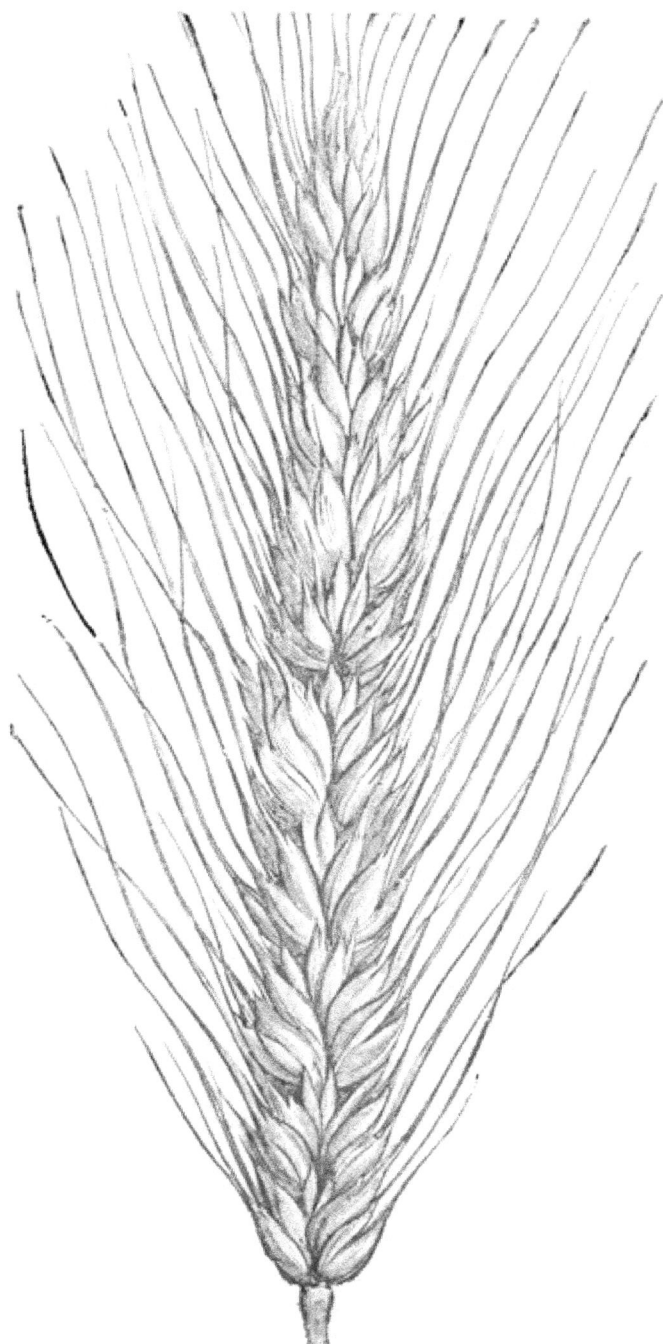

Fig. 4.—CHAMPLAIN WHEAT.

Fig. 5.—DEFIANCE WHEAT.

bushel, one thousand three hundred and forty pounds of straw, and two hundred and ninety-seven pounds chaff."

In Wisconsin, as in most of the prairie States, spring varieties are suited to large portions of the State. In spring wheats, Arnautka, Club, Odessa, Fife, and Russian White, are most popular; in the winter wheats, Clawson, Genesee Flint, Gold Dust, Fultz, Jennings, Lancaster, Mediterranean, and Red, are the popular varieties.

THREE NEW VARIETIES.

Recently two new varieties of Spring Wheat have been produced in Vermont. They are reported as giving large yields, and being valuable, and are represented in the engravings figs. 4 and 5 on the previous and this page. The "Champlain" is a bearded, red-kernel wheat; the other, "Defiance," is a white, bald wheat,

Fig. 6.—RUSSIAN SPRING WHEAT.

and is generally preferred on account of its lighter color, and being beardless.

We also give an engraving (fig. 6) of a new Spring Wheat, called the "White Russian" (somewhat like the Defiance), which, it is claimed, is a great cropper, and very valuable.

SOME ENGLISH PEDIGREE WHEATS.

Mr. T. E. Pawlett, an English farmer, reports in detail some interesting experiments. He says: "October twenty-fifth, 1861, I drilled in the following seven sorts of wheat, in drills eight inches apart, covering about one and one-half inch, six pecks the acre, on clover-sod plowed-under, after being fed a short time by sheep, and obtained results as follows:

1—Hallett's Pedigree, red	36½	bushels per acre.	
2—Giant, red	38¼	"	"
3—Tibbald's Wonder, red	43¼	"	"
4—Corner's, white	42¼	"	"
5—Glory of the West, white	37½	"	"
6—Grace's, white	44	"	"
7—Russian, white	41½	"	"

October twenty-sixth, same year, he made another experiment, on another field, with six varieties, on gravel land, after clover plowed-under, and same quantity of seed drilled-in, same distance and depth as in the above experiment, with the following results:

1—Tibbald's Wonder, red	48½	bushels per acre.	
2—Giant, red	38¼	"	"
3—Browick, red	44¼	"	"
4—Russian, white	33½	"	"
5—Corner's, white	45½	"	"
6—Talavera, white	36½	"	"

He remarks that, from these experiments, it appears that Corner's and Grace's are the best yielders of the white wheat, and that the Giant and Tibbald's are the best yielders of the dark wheats, on his land, while Tibbald's gave the heaviest yield of all; Corner's is the best quality of grain.

3

CHAPTER IX.

GREEN MANURING AND PLOWING.

PLOWING-IN GREEN CROPS.

It is, probably, safe to say that no other mode of fertilizing land—either to preserve or restore productiveness—is so effective and cheap as plowing-in green crops, such as clover, lucern, peas, buckwheat, and some others, treated with liberal top-dressings of lime or ashes just before plowing, and with plaster while growing. This practice not only supplies the soil with vegetable matter, but it tends to make it friable and porous, so that the air can permeate freely, and allows the roots of the plants to run and spread freely for their needed nourishment.

It lightens up, leavens the land, as it were, doing much to prevent the evil effects of drouth by creating and preserving a degree of moisture in the soil during a dry time. Lucern, or Alfalfa, as it is called in some sections, is even better than clover in the estimation of those farmers who have used it, as it runs its roots deeper than clover. The roots are also larger, and tend to subsoil culture, and when cut off eight to ten inches deep in the soil by the plow, they leave it moist and porous to that depth while decaying, and make a favorite bed for the roots of the wheat plant.

PLOWING PRAIRIE LAND—THE OLD WAY.

At the time of our first becoming a settler in the Western States, the ordinary mode adopted by the pioneers for "breaking prairies" was with a heavy team—four to six yoke of oxen—and a large "break-up plow" that would turn a shallow furrow, twenty-four to thirty-six inches wide, two to three inches deep, and this broad, thin ribbon-like strip of prairie sod would be laid over smooth and flat as a strip of carpet. The aim was to cut and turn it as thin and wide and flat as possible and have it hang together, and be fairly inverted, each succeeding furrow lying nicely down in the preceding one, so that few spaces would be left for grass to grow up to the surface, with a depth that should be just under the main roots of the grass, generally from two to three inches.

The plowing was generally done in spring or early summer, in order that the vegetable matter might become decayed for sustaining the wheat, corn, or other crop that might be planted upon it; or in the fall, if the settler then first entered upon the land, and thus be ready for an early spring crop, as soon as the frost was out of the way. It was always a pleasant, satisfactory occupation to hold or follow the huge breaking-up plow, drawn steadily along by the stalwart team, as there was always such a sense or feeling of conquest, of subjugation.

PLOWING PRAIRIE LAND—THE PRESENT WAY.

But now these things are somewhat changed, and recently a better and more elaborate mode has been adopted, as thus described in a communication to the AMERICAN AGRICULTURIST, as follows:

"At Schuyler, Nebraska, West of Omaha, J. T. Clarkson showed some fields of prairie prepared for wheat which were broken up by him in the spring; he first turned over the virgin sod, about three inches deep, in the usual way; then a second plow followed in the furrow and took up about an inch more of the soil and threw it over the inverted sod; this, being carefully harrowed, filled up the spaces between the sods and furnished a fine soil seed-bed for the grain."

"At Marshall, Minnesota, E. S. Youmans treated a part of his land thus: He broke it up on May tenth;

July tenth to September fifteenth. The disk harrow or
sod-cutter was used and the sod all cut finely ; it was
then 'back set,' that is, the plow was run under somewhat deeper, and the cut sods were buried under the
loose, turned-up soil. On this seed-bed spring wheat was
sown from the sixth to the twentieth of April. Thus
treated the prairie land will yield five to seven bushels,
per acre, more than with the usual single plowing. W.
L. Nevins had six hundred acres of spring wheat, near
Tracy, Minnesota, five hundred and forty of which were
treated like the above, and it seemed to give a yield of
about eight bushels, per acre, more than that with the
single plowing. He sowed the Fife wheat, from the
sixth to the sixteenth of April, fifty quarts of seed to
the acre."

Double plowing, cutting the sod finely and covering it
with the rich, friable prairie, making a loam-bed of the
whole, was certainly a paying operation.

PLOWING IN THE GULF STATES.

With an improved cultivation, deeper, finer plowing
and pulverization, much more of the lands of Florida
and Georgia can be made to produce good yields of
wheat. But before the deep plowing is done it is necessary to have the land well underdrained to the depth of
at least two feet, in order to secure the advantages of the
deep plowing ; and the plowing should be done with
good, heavy two-horse or three-mule teams, then thoroughly harrowed and rolled, to compcletely pulverize the
land. This treatment will insure a good crop of wheat
on all the ordinarily fair lands of the Gulf States, but
the single-mule plowing, which, we are informed, generally prevails there, will never secure uniformly good
crops of wheat, there or elsewhere. Land must be well
drained and deeply tilled to produce wheat.

CHAPTER X.

RECAPITULATION OF OPERATIONS.

We will here sum up, in brief, the process or requisites essential to produce increased yield of wheat and continued good crops, as follows :

First—PERFECT DRAINAGE, by both under-drains and surface ditches, as shall be found necessary to prevent stagnant water in the sub-soil or any standing water on the surface, for any length of time after the thawing of ice and snow, or after heavy showers.

Second—DEEP CULTIVATION, by sub-soil plowing or trenching, at least twelve to fifteen inches deep, in order that plant roots may run deeply for sustenance, and also that moisture may rise from below to the surface in seasons of drouth.

Third—ALKALINE MATTER.—The soil needs a liberal supply of ashes, lime, or other substances of alkaline properties, and also salt. A two-fold benefit is caused by these ingredients in the soil, namely—they aid largely in dissolving the silicia (or flint) and they are, to a considerable extent, preventives to ravages of insects and of diseases; especially the salt, which is effective, very often, in preventing injury by rust. Any or all of these things are beneficial to the wheat crop, particularly where there is prevailing liability to rust and crinkling straw.

Fourth—CLOVER AND PLASTER ROTATION, the frequent use of, and plowing-under of various green crops as manures : the plaster to be applied to the clover, or other crop to be plowed-under, to induce ranker growth, together with the liberal application of lime to the land by being harrowed into the surface before seeding.

FIFTH—THE SEED.—Careful selection of and brining the seed in salt, and drying in lime or plaster.

SIXTH—HARROWING AND ROLLING.—The land, just before seeding with the drill, should be thoroughly harrowed and rolled, to crush all lumps and completely powder the soil, so that the largest possible portion of it will be available to nourish the young plants. Another object is to make a soft, mellow seed-bed into which the drill can drop the wheat, and have fine earth to fall back into the drill furrows to cover the grain perfectly at even depth, with no hard, coarse lumps to hinder or smother the growth of the young wheat.

SEVENTH—HOEING OR CULTIVATING the growing wheat in fall and spring, often enough to keep down weeds and keep the soil mellow and moist, which will greatly increase healthy growth, letting in air and sunshine more freely, and will also facilitate the applying of remedies for diseases, as well as the dislodging of insects when they infest the crop.

EIGHTH—EARLY HARVESTING—Much will be added to quantity, quality, and safety of the crop by early harvesting, while the wheat is in the soft, dough state, which tends to prevent injury by rust, loss by shelling and bad weather ; enables the work to be better done by not crowding so much into a short space of time, and the work is more pleasant, as the straw is softer and tougher ; furthermore, as has been shown in previous pages, early harvest makes heavier grain, while the same weight of grain makes more and better flour.

MORE KNOWLEDGE NEEDED.

No matter how much a farmer may know or ave learned by reading, or from experiments made by his neighbors, he can be further enlightened and benefited by making experiments himself on questionable points, or in regard to practices of the utility of which he is not assured. He can make the experiments at first on a small scale, if he wish, so that the loss will not be great or disastrous, in case of failure.

More scientific and practical knowledge would enhance both the pleasure and profits of agriculture, were the large mass of farmers better informed in regard to Botany, Chemistry, Geology, Mineralogy, and the Physiology of Animal and Vegetable Life, it would be greatly to their advantage, by enabling them to make their farm operations both more effective and productive. For this reason practical agriculture should be taught as a regular study, by competent teachers, in all of our district and academic schools.

Large numbers of the children, especially in the rural schools, are to grow up practical farmers, and they should be armed and qualified as thoroughly as possible, with such education and knowledge as will prove of advantage to them in their special avocation, and render them as useful and intelligent citizens and farmers as they are capable of becoming; and they should receive the rudiments and first principles of such education when young and in the primary schools.

Dr. Blake, the distinguished scientist and educator, once said in an address, that "Lecturers, in all parts of the country, should be sent out and maintained by the Government, and the farmers should hear them every month on topics interesting to them as cultivators and stock breeders—lecturers of ability and learning."

CHAPTER XI.

EXAMPLES OF SUCCESSFUL WHEAT CULTURE.

As an encouragement, especially to our younger farmers, and as a stimulant to all, to make efforts for the highest possible achievements in wheat growing, we present many examples of large yields per acre from various sections of our country by different farmers, who have far exceeded the common yield of thirteen to fifteen bushels, which has been the average throughout the country for several years past. While in Massachusetts, Michigan, Rhode Island, and Oregon, the average, per acre, in 1878 and 1879, was about twenty-two bushels; in Illinois, New York, and Ohio, it was nineteen; California, Kansas, Indiana, Texas, and Vermont, seventeen; Pennsylvania and New Jersey, fifteen; in all of the other States, as low as fourteen or under, and in some of the States as low as six to eight bushels.

Now, we believe the lowest of these named may easily reach the figure of the highest, and that many of the States may attain an average of thirty to forty bushels to the acre, simply by fairly adopting the thorough system and methods pointed out in these pages.

One farmer, of Hudson, Ohio, stated in the " Country Gentleman," that he got from a field of sandy-clay loam land, thirty-two bushels of Clawson wheat, and twenty-four bushels of Fultz, per acre; that he weighed in the scales kernels of each, and found that thirty kernels of the Clawson balanced forty kernels of the Fultz, and that he planted eight pecks of Clawson and seven pecks of Fultz to the acre. On that portion of his land which

was not well under-drained, both varieties suffered somewhat by winter killing; otherwise his whole yield would
have been one-quarter larger, while no injury occurred
from that cause on the well-drained land; the largest
yield he ever knew from the Fultz was forty bushels the
acre, while his best three acres of Clawson gave one
hundred and eighty-one bushels, being sixty and one-
third bushels per acre."

Mr. Harroon, of Monroe County, N. Y., obtained from
eleven and three-quarter acres of clover turned under,
four hundred and forty-three and one-half bushels of
handsome Blue-stem wheat, being over thirty-seven
bushels per acre.

Ellwanger & Barry, of Rochester, N. Y., thrashed
from eight acres an average of fifty and a half bushels of
good wheat on land thoroughly drained and well worked,
which had previously been a nursery and orchard, showing the advantage of having land well drained and perfectly pulverized for wheat.

A correspondent of the old "Genesee Farmer" reports
a crop of Genesee Flint wheat giving ninety and three-
quarter bushels on one acre of land, containing, by analysis, only two and forty-three one-hundredths per
cent of organic matter, but contained thirty per cent
(very large) of soluble silicia, with potash, soda, and
other minerals, in larger proportion than is generally
found in good lands.

The "Michigan Homestead" says that Dr. Smith
stated, in an address before the Saganaw (Mich.) Farmers' Club, that David Geddes, of that county, obtained
seventy-three bushels of good wheat from one acre of
land. James L. Rea, of Lewis and Clark County, Mon-
tana Territory, produced one hundred and two bushels of
good wheat from one acre, and he obtained the first pre-
mium, at the Fair, for the largest yield of wheat raised
in the Territory.

OTHER ENCOURAGING EXAMPLES.

It is stated, on what is regarded good authority, that
a farmer in Lake County, Colorado, sowed one acre of
sandy land, May first, with White Russian wheat, and in
September harvested from it one hundred bushels of
good, sound grain. The land was irrigated with water
from a mountain stream.

A farmer in Carroll County, Illinois, reports that for
several consecutive years he obtained twenty-five bushels,
the acre, of Odessa Spring wheat, from the same field ;
he also found that the Odessa answers a good purpose
as a fall wheat, giving that yield, sowed either in fall
or spring, in that region.

Some time since it was reported in the " Ohio Farmer "
that a Mr. Cavin, of Indiana, obtained an average yield
of forty-nine bushels per acre from eleven acres ; also,
that Mr. Richards, of Ohio, obtained nearly the same
average yield from an entire field of twenty-seven acres,
and that Andrew Smith, same State, obtained an average
of fifty-four bushels the acre from fifteen acres, with the
Clawson variety. Mr. French, of Berkshire, Massachusetts, obtained, by drainage and thorough cultivation, an
average of fifty-five bushels the acre, with the Clawson
wheat, one acre of the same field giving sixty-five bushels ; the Clawson is noted as a remarkable tiller, hence
its large yields. Father Weikamp, of the Convent Farm
in Emmet County, Michigan, is reported to have thrashed
one hundred and seventy-four and one-half bushels of
wheat from three and one-half acres of land, giving a
fraction over fifty bushels the acre. A Bel Air (Md.) paper states that William Oldfield, of that county, in 1878,
raised one thousand bushels of wheat from twenty-eight
acres. Part was sown with Fultz wheat, giving forty-five
bushels the acre. The balance was sown with Mediterranean, which gave thirty-five bushels the acre. One

grower in Arkansas reports getting eighty-two stalks, in one stool, from one kernel of Fultz wheat.

YIELD AND PRODUCT FOR SIXTEEN YEARS.

From statistics in the Agricultural Reports, for the fifteen years previous to 1878, it is learned that the total average area sown of wheat was twenty million five hundred and seventy-nine thousand eight hundred and thirty-one acres ; total average product, two hundred and fifty million two hundred and seventy thousand one hundred and twenty-seven. In 1863, thirteen million ninety-eight thousand nine hundred and thirty-six acres were sown, producing one hundred and seventy-three million six hundred and seventy-seven thousand nine hundred and twenty-eight bushels of wheat, and showing an average yield, per acre, of a fraction above thirteen bushels, for that year. The average yield, per acre, during sixteen years, including 1878, was found to be twelve and one-half bushels. In 1878, the area harvested was reported at thirty-one million acres, and the product at about four hundred and twenty million bushels, giving a fraction over thirteen bushels per acre. The average price, per bushel, for sixteen years, was one dollar and twenty cents and four mills ; average price from 1871 to 1878, inclusive, was one dollar and four cents ; the highest average price, any one year, during the sixteen years past, was two dollars and six cents and four mills. When the writer was a boy, on the Genesee Flats, fifty years ago, it was a common thing among farmers to obtain as high as forty, fifty, and often sixty, bushels the acre.

RESPONSES TO MY CIRCULARS.

During the latter part of last year I sent out several hundred circulars to reliable and practical parties, in most of the States, for the purpose of obtaining reports of the best achievements known in wheat growing, by the best and most successful farmers, asking answers to the following questions :

QUESTIONS CONTAINED IN THE CIRCULAR.

What is the largest yield of wheat, per acre, known to you, in your neighborhood, on not less than two acres? On what kind of soil? What the plowing? What the variety of wheat? What date, and manner sown? What date harvested? What the fertilizers used? and other useful information.

Many responses to my circulars, with the desired information, have kindly been returned to me from all parts of the country, showing that some growers, almost everywhere, have succeeded in getting extra large yields, ranging from thirty up to sixty-one bushels to the acre, on the whole of large fields, and portions of many of those replies are given in pages further on.

Chief among the valuable lessons which the reader may learn from these reports, is—that the larger yields advocated by me in this little work, are perfectly and easily practicable for all farmers who possess the ambition and energy to secure them.

TABULATED STATEMENT OF RESULTS IN THE REPLIES TO MY CIRCULARS, FOR THE CROPS HARVESTED, 1879.

Name of Producer.	Post Office Address.	Yield, per acre, bushels.	Kind of Land.	How fertilized.	Depth plowed.	Variety of Seed.	Date of Planting.	Date of Harvesting.	How Planted.
C. H. Dann	Warsaw, N. Y.	46	Clay loam	Stable manure.	Deep.	Clawson.	Sept. 1	July 10	Drill.
N. W. Dean	Madison, Wis.	44	Rich loam.	None.	Medium.	Red Winter.	Sept. 10	July 7	do.
A. E. Blount	Fort Collins, Col.	59	Sandy loam.	None.	Shallow.	Australian.	Sept. 10	July 14	Broadcast.
John McClellan	Minneapolis, Kans.	61½	Prairie.	None.	Deep.	Red Winter.	Mar. 15	July 14	Broadcast.
Ed. Short	Salina, Kansas.	56	Bottom land.	None.	Medium.				do.
E. L. Russell	Tecumseh, Mich.	38	Sand and clay loam.	Clover sod.	8 inches.	Fultz.	Sept. 10	July 15	do.
do.	do.	36	do.	Wheat stubble.	8 inches.	do.	Sept. 11	July 20	do.
do.	do.	36	do.	Wheat stubble.	8 inches.	do.	Sept. 13	July 15	do.
do.	do.	42½	do.	Clover sod.	7 inches.	Delhi.	Sept. 14	July 21	do.
B. J. Bidwell	do.	37½	Gravel loam.		7 inches.	Victor.	Sept. 10	July 15	do.
do.	do.	35	do.		8 inches.	Fultz.	Sept. 12	July 17	do.
do.	do.	36	do.	Wheat stubble.	8 inches.	Clawson.	Sept. 10	July 10	Broadcast.
do.	do.		do.	do.	do.	do.	Sept. 10	July 15	do.
Amos Hollenbeck	Franklin, Mich.	35	Sand and gravel.	do.	do.	do.	Sept. 10	July 10	do.
D. P. Anthony	Lutsen, Mich.	47	Clay loam.	do.	do.	Delhi.	Sept. 10	July 15	do.
Geo. Lester	do.	42	Sand, clay loam.	Stable manure.	6 inches.	Blue-stem.	Oct.	June 28	Plowed-in.
H. C. Hallowell	Sandy Spring, Md.	30	Limestone.		Medium.	Fultz.	Sept. 16	July 10	Drill.
O. N. Bryan	Marshall Hall, Md.	31	Sandy loam.	None.	1 inch.	Red Winter.	Sept. 16	July 18	Broadcast.
D. Lawrence	Howard Co., Md.	30½	Clay loam.	Stable manure.	Deep.	do.	Sept. 10	July 10	Drill.
Lewis Neiman	York, Pa.	54	Sandy loam.	Corn stalks.	5 inches.	Fultz.	Sept. 16	July 12	do.
Luke Eger	Lycoming, Pa.	27	Clay loam.	Stable manure.	5 inches.	do.	Nov. 1	July 7	do.
Jos. Penrose	Coatsville, Pa.		do.	Clover sod.	Deep.	do.	Sept. 15	June 20	do.
Ethan Allin	Pomfret, Conn.	40	Sandy loam.	Clover sod.	do.	do.	Sept. 15	June 25	do.
Wm. Denman	Wakeman, Ohio.	40½	Stable manure.	do.	do.	do.	Sept. 10	July 5	do.
V. C. Stiers	Haydenville, Ohio.	36	Sandy loam.	do.	do.	do.	Sept. 10	June 30	do.
C. E. Thorne	Columbus, Ohio.	28	Clay loam.	Potash, phosphate.	6 inches.	Lancaster.	Oct. 1	June 30	do.
do.	do.	40	do.	Stable manure.	7 inches.	Boughton.	Oct. 1	June 30	do.
Far. & Mech. Bank	Creoline, Ohio.	40	do.	do.	12 inches.	Velvet-chaff.	Sept.	June 25	do.
do.	Canton, Ohio.	46½	do.	do.	8 inches.	Golden-straw.	Sept.	June 28	do.
Henry McDowell	Columbus, Ohio.	33	do.	do.	Deep.	Fultz.	Sept. 12	July 10	do.
James H. Hess	Allamakee, Iowa.	31½	Heavy loam.	do.	Medium.	Gold Medal.	Aug. 26	Aug. 24	Broadleaf.
D. W. Adams	Raleigh, N. C.	40	Clay loam.	Clover sod.	6 inches.	do.	Nov.	June	do.
L. L. Polk	Dunnettn, Ind.	31½	Gravel loam.	Stable manure.	Medium.	Clawson.	Sept. 14	July 14	do.
W. P. Bundy	Haw River, N. C.	45	Sand, clay loam.	Manure, lime.	8 inches.	Virginia White.	Oct. 8	June 28	do.
T. M. Holt	Homer, N. Y.	27½	do. new.	None.	5 inches.	Red Winter.	Sept. 13	July 15	Broadcast.
Spencer D. Short	Wilmington, Del.	40	Black loam.	None.	12 inches.	Russian.	Oct. 13	June 26	Drill.
T. T. Budd	Fort Atkinson, Wis.	49	Clay, sand loam.	Stable manure.	8 inches.	Clawson.	Sept. 4	July 4	do.
R. S. White	St. Joseph, Mo.	30	Clay, gravel loam.	None.	10 inches.	Clawson.	Sept. 20	July 20	Broadcast.
W. H. Concord	Charlton, N. Y.	30	Gravel loam.	Phosphate.	10 inches.	Clawson.	Sept. 15	July 10	do.
F. D. Curtis	Cleveland, N. Y.	60	Clay loam.	Clover sod.	9 inches.	Clawson & Wicks.	Sept. 10	July 10	Drill.
Richard Johnson	Fentonville, Mich.	32	Clay loam.	Clover sod.	9 inches.	Clawson.	Sept. 10	July 20	Drill.
George Judson	Centreville, Mich.	44	Sandy loam.	Mint sod.	7 inches.	Early May.	Sept. 10	July 8	do.

CHAPTER XII.

EXTRACTS FROM LETTERS.

Agricultural Commissioner L. L. Polk, of North Carolina, writes : "We have many other as large yields of
wheat as the one reported here (thirty-one and one-half
bushels), but not on such large areas of ground as the
above ; deep plowing and fine pulverization does it."

Mr. V. C. Stiers, of Ohio, says : "Fultz is their best
variety of wheat ; Dr. Little gets large yield by hauling
the dead animals and other stuff from the town, and
then composting them with the manure and garden
earth on his farm ; it gives him very profitable returns
for the cost."

Mr. D. Lawrence, of Maryland, says : "The land
was well prepared by harrowing and rolling before seeding with the drill ; the seed was carefully screened and
brined, to make it perfectly clean."

Mr. J. H. Hess, of Ohio, says : "My yield of Ar-
nold's Gold Medal, in 1878, was forty-five bushels the
acre ; in 1879, on similar soil, it is forty-six and one-half
bushels the acre."

Hon. Richard Johnson, of Livingston County, New
York, writes : "When this county was new, half a cen-
tury ago, yields of wheat as high as sixty bushels the
acre were raised here in some instances, and often forty
bushels. Now, we think twenty to twenty-five bushels
the acre a good yield ; our land, generally, is a gravelly
loam. The reason that we do not get such crops as formerly is that the farmers "run" their land too much
with grain, and do not pasture and clover enough ; and
the forests are cut away, so there are no trees for wind-
breaks."

Mr. Adam Bloom, of Michigan, says "he plowed-under an old mint-stubble, in the spring, about seven inches deep; dragged it twice, and cultivated it four times in all, with a good cultivator, wherever the mint and grass made their appearance, so that the ground was made fine and kept clean as a garden all the season, until seeding time; then cultivated about five inches deep so as to bring the rotted mint manure near the surface; then planted the wheat with a drill. This mode gave me eighty-eight bushels of good wheat, Early May variety, from two acres, on an old mint stubble, well cultivated and cleanly subdued."

Mr. E. L. Russell, of Mich., says his ground was clover sod, plowed-under the year before, and then in August wheat stubble was plowed about eight inches deep; then, early in September, just before drilling in the seed, eighteen loads of barn-yard manure to the acre were spread on the plowed ground and thoroughly harrowed into the surface soil, and followed with the roller; then, September eleventh, planted by drill, putting one and a half bushels of seed to the acre.

Such statements, from practical farmers and successful growers, are very valuable.

CHAPTER XIII.

DISEASES AND INSECTS ATTACKING WHEAT.

Many of the diseases to which the wheat crop is liable
are caused by improper culture and conditions of the
land, as has been shown in the foregoing pages. Some
experienced growers maintain that even the prevalence of
insects may be prevented by judicious culture of the soil
and preparation of the seed.

RUST AND SMUT.

Rust, smut, and other forms of fungus, are usually
due to a lack of drainage, stagnant water in the sub-soil,
and a too succulent growth of the plant by the excess of
nitrogenous matter, and a lack of soluble silica in the
soil, which cause soft, spongy straw, not sufficiently
glazed over with silica to render it hard and stiff, to resist effect of changes in temperature.

It is maintained that perfect drainage and complete
pulverization of the soil, so as to freely admit the circula-
tion and action of the air and moisture all through it, by
which silica and other mineral matters will be better dissolved, will almost entirely remedy the evil, especially
with a liberal quantity of potash and salt in the soil.
Ashes, lime, and salt, with moisture, are powerful solvents of all matters in the earth necessary to make stout,
healthy wheat crops. Hence their action does much to
prevent rust and smut, as also does soaking the seed, six
to ten hours, in salt or blue-stone brine, and then stirring
the grain in lime or plaster, liberally, to dry it, for working freely through the drill in planting.

A moderate dressing of lime on the growing wheat,
late in fall or early spring, when wet with dew or rain, is

a good preventive and cure. But early harvest, while the grain is soft, is a very sure way to avoid the destruction by rust.

THE WINTER KILLING OF WHEAT.

Winter killing is almost entirely prevented by having the land thoroughly drained, two to three feet deep, and deeply cultivated to the depth of twelve or fifteen inches. It is also prevented, to a great extent, by having the soil completely pulverized by harrow and roller, so as to enable the drill to make little ridges and perfectly cover the seed, leaving it in a shape that the wind will not blow it bare in winter or spring, but will, in fact, cover it still better, the ridges holding the soil.

THE CHINCH BUG AND HESSIAN FLY.

In some sections and seasons the Chinch Bug (*Micropus leucopterus*) is very destructive, especially in dry seasons, but wet weather is unfavorable to it ; all grains are more or less liable to be infested by this insect. Experienced farmers have found that spreading eight to ten bushels of quick lime, to the acre, on the stubble and among the weeds, and plowing it all under in August, for seeding in September, is a pretty sure way of getting rid of this pest, as well as many other insects which infest the wheat crop ; but a second shallower plowing, or working with the cultivator, and thorough rolling, before seeding, should be done, to fit the land nicely for the drill, and to more perfectly mix the lime and soil.

Another troublesome pest is the Hessian Fly (*Cecidomyia destructor*) which often appears in some localities and seasons.

A writer in the "Allentown (Penn.) Democrat" says :—"There are two broods of the Hessian fly brought to perfection each year—in the fall and the spring ; the transformation of some appear to be often retarded beyond the usual time, and the life of individuals is sometimes longer than a year, and the continuation of the species in after years made sure. The mature insect deposits its eggs on the young plants soon after they appear above the ground, and are several weeks doing this ; the eggs are about five days in hatching, the young worms going directly to a joint of the stalk, where they affix themselves and become stationary, until their transformations are completed, but do not go to the center of the stalk, nor bore into it, as some suppose, but lie upon its surface, protected by the leaves. One maggot seldom destroys a plant, but three or four deplete it of its juices, and it dies. It takes five or six weeks for the larvæ to attain full size. At this time the skin hardens, becomes brown, and to the naked eye the insect has the appearance of a small flaxseed. In this condition it remains until spring, when the fly comes forth, and goes through the same operations as before."

A dressing of three to five bushels of salt, to the acre, in the fall, and another in early spring, it is said, will effectually destroy them ; and that lightly covering the seed in rich, friable soil, is more unfavorable to their growth than the opposite. A Virginia farmer recommends the sowing on wheat of four to five bushels of lime, to the acre, as a remedy for the Hessian fly. Sow while the dew is on the plant, and the lime will be dissolved, forming a lye, which runs down the blade to the root, thus destroying the insect.

Plaster sowed on the growing crop, spring and fall, is said to be very useful.

WHEAT MIDGES.

Mr. Klippart speaks of two insects known as Midges— the red midge (*Cecidomyia tritici*), a species of the same genus as the Hessian fly ; and another, the yellow midge, a small fly. They both prey upon the head of the wheat, in the chaff, and on the kernel, while the grain is green, and cause it to blast before coming to maturity.

Rich culture, strong growth, with early planting and early harvesting, will do much to prevent the evils done by these insects. Wheat sown in wide drills and hoed, by hand or horse-hoes, gives a favorable opportunity to apply lime or sulphur to the heads of the grain, by sprinkling, which cannot be done in ordinary culture; and those articles are destructive to the midges, if applied when the grain is wet with dew, or rain, to retain the sulphur a longer time.

THE GRANARY OR BARN WEEVIL.

The Weevil (*Calandra granaria*) is an insect which infests grain in the granary. The weevils prey upon all kinds of grain in the bin and the corn-crib, and being very small, about one-eighth of an inch in length, they are not readily seen, particularly in a dark bin. Their mode of mischief is by piercing minute holes in the kernel, and there depositing their eggs, from which are hatched small maggots which eat out the heart and flour of the grain.

An agricultural journal remarks in regard to it: "Wheat in the granary is subject to injury by the weevil and the grain moth. This damage may be prevented, to some extent, by shifting the grain and running it through the fanning-mill. Corn cribs are almost always infested by rats and mice. A vermin-proof crib may be made by covering the posts and lower corners with tin or sheet-iron, which may be painted for preservation. The loss by these causes will average eighteen per cent, and often more, of the value of the grain, but it may be in part or wholly avoided by care and precaution." Fumigation with sulphur or tobacco has been found useful.

But, as has been and is maintained by many old prac-

tical growers, deep, rich, thorough cultivation of the soil,
with care in selecting and preparing the seed, is largely a
security from serious injury by any of those diseases and
insects, by producing plants with vigor and strength to
resist or overcome ravages by either. Slim growth and
feeble conditions induce and invite ravages by disease and
insects, while luxuriant growth and healthy conditions of
soil are, as a rule, favorable to security.

CHAPTER XIV.

IMPROVED MACHINERY AND IMPLEMENTS.

So great and remarkable have been the inventions and
improvements in the implements and tools used by farmers during the last forty or fifty years, rendering their
work so much more easy and expeditious that the farmer
has or might have much more time to read and study
all subjects of interest to his profession ; and hence we
would reasonably conclude that his operations would be
more productive, that yields would be larger, and his soil
be better preserved from exhaustion ; but, unfortunately,
we do not find the facts showing such to be the case—at
least not in the matter of higher yields and preservation
of the soil, for the opposite is the fact. The general yield
per acre is less than it was forty or fifty years ago, and
lands are much "run down."

New inventions have enabled the producers to draw from
the soil its powers and productions with such rapidity,
without equally replenishing its fertility, that the capac-
ity of the land to produce has been almost as rapidly ex-
hausted ; whereas, had the farmers as generally applied

those vast powers also to draining and thoroughly pul-
verizing the land to an additional depth, the fertility or
productive power would have been proportionally increased and preserved.   It is not yet too late, if they
will learn lessons of wisdom and judiciously apply them
to a more perfect system of tillage.

Land which is lumpy and cloddy, only partly crushed
and mellow, can be only partly available to nourishing
and maturing of the crops, as plants can appropriate
only what is fit for solution ; for continuous, large cropping this should be done, and machine-power can well
be adapted to do it.   So with deep culture ; land thoroughly and uniformly cultivated to the depth of twelve
or fifteen inches is capable of producing, year after year,
nearly twice as much crop as six inches in depth ; and
machine-power, which has been so effective in harvesting
and thrashing, can be made equally effective in draining
and comminuting the land to greater depth.   Then the
rapid cropping will not exhaust the power of production
of the soil.

CHAPTER XV.

ANALYSES OF WHEAT AND STRAW.

Knowing the constituents of the wheat plant—both grain and straw—will aid very much in determining what parts are derived from the earth and what from the atmosphere, respectively, as well as what is to be done, by cultivation, to supply those wants so as to secure the best results.

Burning the grain and straw, reducing them to ashes, shows the mineral or inorganic ingredients which are obtained from the soil; chemical decomposition and separation shows the organic and nutritive ingredients, which are mostly obtained from the atmosphere.

Vœlcker proves that a large percentage of wheat is starch, gluten, and sugar, while straw contains a large per cent of carbon and silica.

Analysis by Boussingault shows that wheat contains, in one hundred parts : Carbon, 46.10 ; oxygen, 43.40 ; hydrogen, 5.80 ; nitrogen, 2.30 ; ash, 2.40=100. These are derived from both air and soil, but mostly from the air.

Prof. Way gives the following analyses of the ashes of grain and straw, separately, in one hundred parts : GRAIN—Silica, 5.63 ; phosphoric acid, 43.98 ; sulphuric acid, 0.21 ; lime. 1.80 ; magnesia, 11.69 ; peroxide of iron, 0.29 ; potash, 34.51 ; soda, 1.87 ; loss, 0.02. Of STRAW—Silica, 69.36 ; phosphoric acid, 5.24 ; sulphuric acid, 4.45 ; lime, 6.96 ; magnesia, 1.45 ; peroxide of iron, 0.29 ; potash, 11.79 ; soda, none ; loss, 0.46. Much silica in the straw, far more than in the grain, and is derived from the soil.

In "Encyclopædia of Agriculture" we find the follow-

ing analysis, by Prof. Beck, of the constituents of grain :
Water 14.0 ; gluten and albumen, 14.6 ; starch, 59.7 ;
gum and sugar, 7.2 ; cellular and woody fibre, 1.7 ; fatty
matter, 1.2 : mineral matters, 1.6=100.

Professor Horsford, in his excellent work on the Paris
and Vienna Expositions, gives the following analyses of
the ash of average good wheat : potash, 30.00 : soda,
3.50 ; magnesia, 11.00 ; lime, 3.50 ; oxide of iron,
1.00 ; chloride of sodium, 0.50 ; sulphuric acid, 0.50 ;
phosphoric acid, 46.50 ; silica, 3.50=100. He gives the
following organic and nutritive ingredients :—Starch,
57.00 ; dextrine, 4.50 ; fibrine, 9.27 ; nitrogen, 2.23 ;
oil, 1.80 ; woody fibre, 6.10 ; ash, 1.70 ; extract matter,
1.40 ; water, 16.00.

Professor Muller found the following in 100 parts of
heavy wheat grains :

Water, 15.65 ; woody fibre, 2.54 ; ash, 1.57 ; nitrogenous matter, 11.84 ; oil, 2.61 ; sugar, 1.41 ; starch,
64.38 ; albumen and gluten are included in the above as
nitrogenous matter, and with the starch constitutes the
nutritive matter.

Professor Way showed that an acre of wheat, which
yielded forty bushels, gave in weight :

Grain, 2,604 lbs. ; mineral matter, 44'/₂ lbs. Straw,
2,775 lbs. ; mineral matter, 123'/₄ lbs. Chaff, 401 lbs. ;
mineral matter, 47'/₂ lbs.

The grain gave 5.6 per cent of silica, and the straw
gave 69.36 per cent of silica ; the grain gave 43.98
per cent of phosphatic matter, and the straw 6.24 per
cent of same. Of lime, potash, magnesia, and soda, the
grain gave 49.87 per cent, while the straw gave only
20.20 per cent of the same ingredients. This large
amount of silica (dissolved sand) in straw and chaff
should go back to the soil for the benefit of future crops.

CHAPTER XVI.

CONCLUSION.

When all the States east of the Mississippi River bring their wheat yield up to that of Michigan and Ohio, the center of wheat production will continue to be east of that river ; but at present, indications are that the wheat center is rapidly tending to a line west of that, if, indeed, it be not already beyond it, so that unless the eastern portions of the country do not speedily improve their modes of cultivation and increase their yield of wheat, they will soon and surely lose their ascendancy in wheat measures.

Better drainage, deeper plowing, and more perfect pulverization of the soil are absolutely necessary, together with a more liberal use of Clover, Plaster, and Lime, to secure a considerably larger yield of wheat, in the older States. Hon. Thomas Pollard, of Virginia, says clay land, with clover fallow, will bear one hundred bushels of lime to the acre, with advantage ; but on land deficient in vegetable matter, very much less should be used.

Highest success in wheat-growing involves and presumes skillful and intelligent management in other parts of farming, so that he who uniformly secures superior results with wheat, and does not impoverish his land or soil, cannot well be other than a good farmer, able to secure profitable results in all other farm operations. Hence, to become an eminent wheat-grower is to become a complete farmer. To aid in bringing about that result is our aim and purpose in writing this little work.

Gardening for Profit:

A GUIDE TO THE SUCCESSFUL CULTIVATION OF THE

Market and Family Garden.

By PETER HENDERSON,

Author of "Practical Floriculture" and "Gardening for Pleasure."

FINELY ILLUSTRATED.

A now well known and standard work on Market and Family Gardening. It is the first book of the kind prepared by a Market Gardener, in this country. The author's successful experience of more than twenty-five years, enables him to give a most valuable record. It is an original and purely American work, and not made up, as books on gardening too often are, by quotations from foreign authors. Everything is made perfectly plain, and the subject treated in all its details, from the selection of the soil to preparing the products for market. Cloth, 12mo.   PRICE, POST-PAID, $1.50.

Gardening for Pleasure:

A GUIDE TO THE AMATEUR IN THE

Fruit, Vegetable, and Flower Garden

WITH FULL DIRECTIONS FOR THE

Greenhouse Conservatory and Window Garden.

By PETER HENDERSON,

Author of "Gardening for Profit" and "Practical Floriculture."

ILLUSTRATED.

This work is prepared to meet the wants of all classes, in Country, City, and Village, who keep a Garden for their own enjoyment rather than for the sale of products. It is adapted to meet the wants of the amateur in in-door and out-door gardening. It is one of the best guides to Window Gardening we know of. The work includes fruit, vegetable, and flower-gardening, greenhouses and graperies, window gardening, and Wardian cases. Cloth, 12mo.

PRICE, POST-PAID, $1.50.

ORANGE JUDD COMPANY, 245 Broadway, New York.

www.ingramcontent.com/pod-product-compliance
Lightning Source LLC
Chambersburg PA
CBHW021959190326
41519CB00010B/1331